AF518628

AGROCHEMICALS
Hazard and Toxins

AGROCHEMICALS
Hazard and Toxins

by
Bhagyashree Chauhan | Tejas Makavan | Om Bhardwaj

Wordpen Academics
Publishing Science That Matter

© 2025 PUBLISHER

Information contained in this work has been published by Wordpen Academics and has been obtained by its author(s)/editor(s) from sources believed to be reliable and are correct to the best of their knowledge. However, the publisher and its author(s) make no representation of warranties with respect of accuracy or completeness of the contents of this book, and shall in no event be liable for any errors, omissions or damages arising out of use of this information and specifically disclaim any implied warranties or merchantability or fitness for any particular purpose.

All rights reserved. Including the right to translate or to reproduce this book or parts thereof except for brief quotations in critical reviews.

ISBN: 978-93-6084-537-7

Published by : **WORDPEN ACADEMICS**
4736/23, Ansari Road, Darya Ganj,
New Delhi – 110 002
Phone: +91-011-35004336
E-mail: info@wordpenacademics.com

Printed at **:** **Replika Press Pvt. Ltd.**

Exclusive Distributor: ***Bio-Green Books, New Delhi***

Preface

"Agrochemicals: Hazards and Toxins" confronts one of the most pressing challenges facing modern agriculture: the delicate balance between agricultural productivity and environmental safety. This book stands as a comprehensive examination of the hazards and toxins associated with the use of agrochemicals, offering invaluable insights into their impact on ecosystems, human health, and sustainable food production.

In an era characterized by burgeoning global populations and escalating food demands, agrochemicals have emerged as indispensable tools in modern agriculture, facilitating increased yields, pest management, and weed control. However, the indiscriminate use and mismanagement of these chemicals have raised profound concerns regarding their adverse effects on environmental quality, biodiversity, and public health.

As editors of this volume, we have endeavored to assemble a diverse array of perspectives from esteemed researchers, scientists, and practitioners, each contributing their expertise to shed light on the multifaceted dimensions of agrochemical hazards and toxins. Through a rigorous examination of pesticide residues, soil contamination, water pollution, and the emergence of pesticide resistance, this book offers a comprehensive understanding of the ecological and human health risks posed by agrochemicals across the agricultural landscape.

Our overarching aim in presenting this book is to foster a nuanced dialogue on the complex interplay between agricultural intensification and environmental sustainability, challenging conventional paradigms and advocating for evidence-based strategies to minimize the adverse impacts of agrochemicals while maximizing their benefits.

We extend our deepest appreciation to all the contributors who have dedicated their time, expertise, and scholarly rigor to this endeavor. It is our sincere hope that this volume will serve as a catalyst for informed decision-making, innovative research, and collaborative action towards a more sustainable and resilient agricultural future.

Bhagyashree Chauhan | Tejas Makavan | Om Bhardwaj

Contents

1

Environmental Risk Assessment of Agrochemicals - A Critical Appraisal of Current Approaches

1. Introduction

The enormous amount of chemical contaminants that are currently found in air, soil, water and sediments of our planet, mostly as a result of human activities [1], call for an evaluation of their risks to the web of life. The contaminants that potentially result in adverse biological effects – whether at individual, population, community or ecosystem level – the so-called pollutants, are of concern [2]. For obvious reasons, pollutants that relate to our food resources are of particular interest and require priority in risk assessment. Among the latter chemicals are those used in agricultural production, also called agrochemicals, which include pesticides of various kinds, plant growth regulators, repellents, attractants and fertilisers. These chemicals are intentionally added to the environment for controlling pests and diseases of crops, requiring an accurate assessment of the maximum amounts that can be applied.

Because of their huge market in all continents, agrochemicals are among the most common pollutants in one fifth of the Earth's land [3]. Also, given the high toxicity of most insecticides, herbicides and fungicides to their target and nontarget organisms, their negative impacts on the environment cannot and must not be ignored. Indeed, pesticides constitute one of the main drivers of population decline in some wildlife species [4]. Therefore, proper evaluations of the risk that pesticides and other agrochemicals pose to organisms and ecosystems should be conducted in a scientific manner.

Various approaches have been used in the past to evaluate the ecological risks of agrochemicals, and all of them focus on the intrinsic toxicity they may have – their poisonous nature – as well as their exposure in the environment. Indeed, the ability of a chemical to adversely affect the functioning of organisms is a major factor to take into account in evaluations of risk. That ability is governed primarily by the specific mode of action of the chemical under consideration, while the magnitude of adverse effects primarily depends on the exposure level and often on the exposure time as well. Here we have the basic components of any chemical risk assessment: toxicity and exposure [5].

Organisms are not only subject to the effects of single, individual chemicals, since the environment they live in is usually contaminated with many compounds, each in different concentrations and with different modes of action. Site-specific risk assessments, therefore, must refer to chemical mixtures if evaluations are to be meaningful. Despite the many advances in this area of environmental science, our ability to properly assess the risk of individual contaminants is still far from desirable, let alone the assessment of chemical mixtures! Even if the characterisation of a pollutant can be done on the basis of its physicochemical properties and toxic mode of action, the exposure component is often difficult to assess, as it involves issues of bioavailability, interaction of the chemical with other natural stressors (e.g. temperature, pH, dissolved and suspended material) and acclimation and adaptation of individual organisms, all of which are difficult to gauge. It is often said that chemical risk assessments lack realism, as they are too reliant on toxicity tests carried out in laboratories, whereas the conditions of the natural environment frequently reduce both exposure and availability considerably [6].

This chapter deals with the ecological risk assessment (ERA) of agrochemicals, deliberately avoiding any discussion about the risk of human exposure to residues of these contaminants in food, which is treated in a different chapter. Emphasis is placed on the limitations of the current approaches, with some suggestions being proposed to improve the evaluations or to overcome specific problems. A critical approach will thus provide insights into the difficulties and challenges of performing adequate risk assessments.

2. Types of Risk Assessment

Chemical risks refer to specific organisms and situations. Furthermore, the primary purpose of their evaluation determines the type of assessment that is carried out. On this basis, two types of risk assessments are usually performed with agrochemicals: (i) ecological risk assessments for protection of ecosystems and (ii) risk assessments for registration of agrochemicals or regulatory purposes, which are partly guided by the need to protect the environment.

Ecological risk assessment aims at protecting organisms, communities and ecosystems. It applies to all man-made chemicals and their degradation products as well as natural contaminants that are poisonous or can be toxic above certain threshold concentrations (e.g. copper, trace metals, nitrate, etc). Usually ERA is site specific, as it considers environmental conditions specific for an area or region in addition to the characteristics and levels of the agrochemicals applied. As a result, risk management plans can be proposed, i.e. remediation processes to reduce or eliminate residues or mitigation measures that reduce the loads discharged into the environment.

Agricultural products that are released into the environment need to be assessed for impacts on ecosystems and consequently be approved by regulatory authorities. The aim of the regulatory risk assessment is to ensure that those chemicals that enter the environment do not pose a serious threat to living organisms. Regulatory risk assessment usually follows preset protocols, such as those of the OECD or US Environmental Protection Agency (EPA). Because any chemical can potentially affect living organisms, regulators can impose restrictions in the usage or establish conditions that mitigate or avoid the exposure of the organisms at risk. While new agricultural products are developed by chemical companies in astonishing numbers, only a fraction of those make it to the market. In the worst case, agrochemicals can be banned and withdrawn from national and/or international markets, as in the case of old organochlorine pesticides (e.g. DDT, dieldrin, aldrin, heptachlor). Obviously, economic considerations should not influence the decisions of regulators in regard to the non-safety of certain chemicals.

3. Framework for Ecological Risk Assessment

Ecological risks of agrochemicals are evaluated by standard procedures, as it is done with other chemicals. This approach is similar to the evaluations of risks for humans, except for one main difference: human risk assessments aim at protecting individuals, whereas ecological risk assessments aim at protecting ecological structures, whether populations, communities or entire ecosystems. Another difference is the preferential use of median effect concentrations (EC50), in particular the median lethal concentrations (LC50) and lethal doses (LD50) in risk assessments of agrochemicals, rather than no-observed-effect concentrations or levels (NOEC or NOEL). This is because the data on the latter metrics are less readily available for the surrogate species used in toxicological testing of agricultural compounds; typically, only mammal toxicity is measured at the lowest levels. Besides this difficulty, NOEC and NOEL are statistically unreliable – it is almost impossible to prove that there are no effects in natural populations of organisms from exposure to low concentrations of chemicals [7, 8].

In essence, all risk assessments rely on the framework shown in Fig. 1, which is based on standard procedures of the US EPA [9] and has been adopted, with some modifications, by most regulatory authorities in OECD countries [10]. The two main components of the risk framework are the toxicity assessment and the exposure assessment [5]. Without toxicity, there is no risk, but if organisms are not exposed to a toxic chemical, they are under no risk either. Therefore, it is only when organisms are exposed to potentially toxic chemicals that a risk assessment must be undertaken.

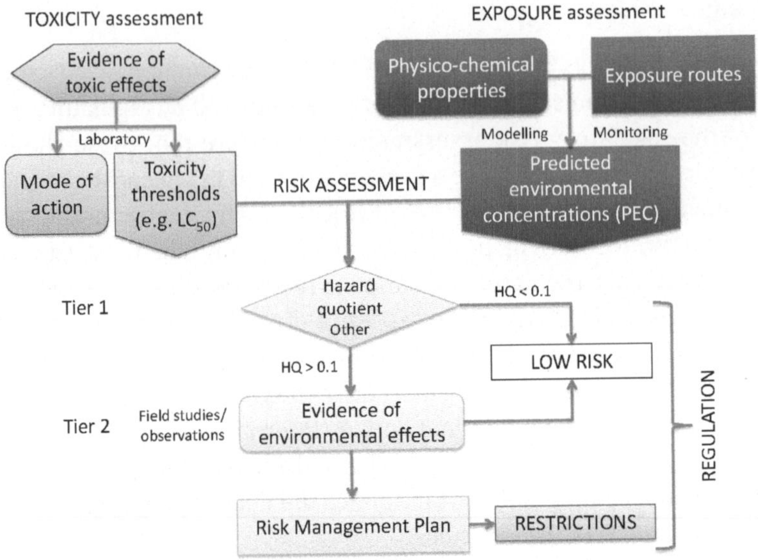

Figure 1. Framework for ecological risk assessment of agrochemicals.

Toxicity levels of a chemical (either concentrations in aquatic media or doses in terrestrial animals) must be determined in the laboratory, as effects from exposure under field conditions may be confounded by other factors. Standard toxicity tests, carried out on several taxa of plants and animals, are primarily designed to establish the acute lethality [11]: either the LC50 for aquatic organisms after a given time of exposure or the LD50 for doses applied to or ingested by terrestrial organisms. In addition, chronic toxicity after prolonged and repeated exposure (whether lethality or another effect) and reproductive endpoints after sublethal exposure over time are often determined as well [12]. Knowledge about the mode of action of the chemical should help understand its toxic effects [13].

The exposure characterisation is complex as it usually involves some sort of modelling to determine the possible concentrations of chemical in the various media in which the organisms live [14]. Inputs for the models include all relevant physicochemical properties of the chemical under consideration, e.g. water solubility, lipophilic behaviour, volatility and degradation constants in water, soil, light conditions, plants and animals. The latter constants are important to understand the persistence of chemical residues in the environment. In site-specific ERA, the exposure should also include monitoring data to validate the modelling used. If monitored and modelled data differ markedly, an explanation should be sought for the discrepancies.

In countries of the OECD, assessment of the two components of risk follows a tiered-structured process (Fig. 1). In the first step, rough estimates of risk are evaluated by a simple ratio, the hazard quotient (HQ) between the predicted environmental concentrations (PECs) in a particular medium and the acute lethality (LC50 or LD50) to the standard test organisms found in that medium. HQ values above 1 are unacceptable, and risk thresholds are commonly set at 0.1, meaning that assessments that result in HQ < 0.1 pass the first tier and can be regarded as "safe" for a particular environment. HQ values > 0.1 require evaluation in a second tier. However, a given chemical may be considered safe to terrestrial organisms such as rats, but not safe for aquatic organisms such as midge larvae. Even within the same environmental medium, some species in the field are more sensitive than the surrogate species used in the laboratory tests. To overcome this problem, probabilistic risk assessments (PRAs) have been derived to identify the fraction of organisms that would be negatively affected by the PEC in a given environmental medium [15]. Risk in a PRA is defined as a feasible detrimental outcome of an activity or chemical, and it is characterised by two quantities: i) the magnitude (severity) of the possible adverse consequence(s) and ii) the likelihood of occurrence of each consequence. Species sensitivity distributions (SSDs) are typically used for that purpose [16] – see below. HQ is the preferred method used in regulatory risk assessments of agrochemicals, whereas PRA is mostly used in site-specific ERA, which in most jurisdictions amounts to a second tier.

Whether HQs or PRAs are used, if the chemical does not pass the first tier, a further evaluation must be done. The second tier involves gathering evidence of effects under realistic environmental conditions. Field trials and mesocosms are typically carried out in order to obtain that information. A number of difficulties with these trials and experimental ecosystems will be addressed in the respective sections below. Suffice to say now that as a result of the second tier, a management plan for the agrochemical under evaluation should be drawn to mitigate its possible risks. Such plans may consider restrictions of usage for a chemical product and other conditions that aim at avoiding the exposure of organisms in the environmental media for which a risk has been identified.

4. The Weight-of-Evidence Approach

Weight of evidence (WOE) is an evaluation of evidence on one side of an issue as compared with the evidence on the other side of the issue or an evaluation of the evidence on multiple issues. WOE is used to describe the type of consideration made in a situation where there is uncertainty and which is used to ascertain whether the evidence or information supporting one side of a cause or argument is greater than that supporting the other side. The US Environmental Protection Agency started to consider a WOE approach for assessment of carcinogenic chemicals in 1986 [17]. The same approach was introduced later for ecological risk assessments [18], and since then, the European Union and other governments have followed suit to address specific problems posed by chemicals. The concept of WOE is familiar to the public, as it is embodied in the justice system of western countries [19]. Several methods, ranging from qualitative to quantitative, are used to reach conclusions from the various lines of evidence synthesised in the WOE approach [20]. In assessing a chemical's safety, however, WOE uses with preference causal criteria and logic methods as well as "probability and other statistical techniques to evaluate how individual data adds to or subtracts from evidence of risk" [21].

It should be realised that, in fact, the WOE approach operates throughout the entire assessment process outlined in Fig. 1. In particular, the toxicity and exposure assessment steps use qualitative and quantitative measures to characterise the chemical in regard to its hazard and potential harm. For example, in the sediment quality "triad", sediment chemistry, toxicity and effects to resident organisms are the individual lines of evidence [2]. In general, individual lines of evidence may include acute toxicity tests, sublethal effects, biomarkers, residue data from field surveys and other sources [5, 22]. The integration of all these lines of evidence is done in the risk assessment step, by looking "critically at each line of evidence and consider uncertainty and variability when individual lines point to different conclusions" [21]. Most ERAs use causal criteria and logic methods to evaluate risks, with sediment assessments relying heavily on the logic methods. The resulting risk is typically defined in terms of probability of harm for populations in a particular environment. The reader is referred to the review by Linkov et al. [21] concerning the application of WOE to current environmental assessments.

5. Toxicity Assessment

The first condition for a risk assessment is to know whether a chemical is toxic or not. Multiple lines of evidence are used for that purpose, including bioassays with standard species that are representative of various environmental media, biomarkers and others.

For decades, acute and chronic bioassays have helped determine the toxicity levels of most agrochemicals to a range of organisms. For all chemicals, lethal toxicity to mammals is based on rats or mice bioassays, from which the oral LD50 and contact LC50 can be determined. In the past, NOEC and the lowest-observable effect concentration (LOEC) were also determined, mainly for human risk assessment. However, ecotoxicologists have been arguing that the latter measures are statistically flawed and consequently should not be used any longer [23, 24]. Instead, the estimation of LD10 or LD20 from standard bioassays is now preferred to define the low levels of effect [25, 26], given that populations of organisms in nature always have a small proportion of casualties due to diseases, lack of fitness, starvation and other factors, which cannot be distinguished from chemical toxicity [8]. Similar bioassays for acute toxicity have been used with a range of organisms, including birds, fish, amphibians, earthworms, springtails, bees, flies and a variety of insect pests, mosquito larvae, midge larvae, dragonfly and mayfly nymphs, several types of crustaceans and zooplankton, clams, oysters, mussels, algae and plant species. The need for standardisation of tests has produced a large body of literature, where some species are preferred based on ecological relevance, sensitivity, commonality, easiness to culture in the laboratory or other convenient traits. The commercial Microtox® assay system is also an acute bioassay designed to detect basal toxicity to the marine luminescent bacteria *Photobacterium phosphoreum*, mostly from some organic substances [27]. Many other standardised *in vitro* cytotoxicity tests have been developed to test a large number of chemicals [28]. Acute lethality is typically determined for single doses at 24 h in terrestrial organisms and after a fixed time of exposure in aquatic organisms: from 24 or 48 h in small organisms to 96 h for larger crustaceans and fish. Chronic toxicity bioassays are carried out to detect possible biological effects after repeated or continuous exposure to sublethal doses or concentrations of a chemical. The endpoints in this case may vary from lethality to carcinogenicity and mutagenicity [29] or from reproduction impairment [30] to hormonal imbalance [31] and developmental effects [32]. Any negative or deleterious effect can be considered a chronic toxicity endpoint, not just the reproduction effects – the criterion for chronic toxicity is long, repeated exposure to chemical amounts that are not lethal within the short timeframe of the acute bioassays. Not all chronic endpoints are considered ecologically relevant, and hence, such endpoints as growth and reproduction would have a higher weighting than others such as biomarkers (unless, of course, these can be linked to community and population effects).

Each one of the above tests is considered a line of evidence for the risk assessment of a chemical. Current information on acute and chronic toxicity for all agrochemicals can be accessed online through the ECOTOX database compiled by the US Environmental Protection Agency (http://cfpub.epa.gov/ecotox/) and other databases with a narrower scope, e.g. AGRITOX (http://www.agritox.anses.fr/index.php), Footprint Database (http://sitem.herts.ac.uk/aeru/iupac/), the Pesticide Manual [33] and others.

Toxic action may depend not only on dose or concentration but also on the duration of exposure. It has long been suggested that toxicity tests should include information not only on doses or concentrations but also on exposure time [34-36], with time-to-effect bioassays being a practical way to achieve this [37]. Toxicokinetics are necessary to interpret the toxicity of a compound with time of exposure [38, 39], because they can relate time-dependent toxicity to target organ concentrations [40]. For some compounds, toxic effects may depend almost exclusively on the total dose [41], whereby the product of dose, or concentration, and exposure time produces the same biological effect (Haber's rule). For pesticides that block specific metabolic or physiological pathways, the toxicodynamics of the compounds determine the time dependency of the effect [42]. In this case, toxicity is better described as a function of the time of exposure in addition to the dose. Understanding of the mode of action of such compounds is, therefore, essential for a correct prediction of their toxic effects [43]. Thus, whenever the interactions of chemicals with critical receptors are either slowly reversible or irreversible and the resulting biological effects are also irreversible (i.e. death), such effects are time cumulative and long lasting [40]. Examples of such chemicals are the neonicotinoid insecticides [44], which impair cognition by blocking nicotinic acetylcholine receptors in the central nervous system of insects and other arthropods. The implications of this time-cumulative toxicity for risk assessment will be dealt with in the section on first-tier assessment below.

One particular group of compounds that have attracted special attention in recent times is the endocrine disruptor chemicals (EDC), among which are some agrochemicals: amitrole, atrazine, DDT, lindane, mancozeb, maneb, metiram, metolachlor, pentachlorophenol, vinclozolin and ziram [45]. EDCs alter the homeostatic balance in organisms by either mimicking the activity of hormones or influencing the metabolism of natural hormones as a side effect. Because hormones are substances that effectively induce or suppress physiological mechanisms, often complex and interlinked, exposure to EDCs can have unforeseen effects, certainly unrelated to the specific mode of action of the ED compound. For example, some organochlorine pesticides induce feminisation in alligators [46, 47]. EDCs should not be confused with hormonal pesticides, which by their very nature mimic natural hormones in plants (e.g. auxin-like herbicides such as 2,4-D) or insects (e.g. ecdysone mimics such as tebufenozide). Identification of EDCs has been given high priority in some developed countries [48] with the aim of restricting their use in view of the potential consequences for human health [49], even if the research needed to characterise and elucidate their impacts is lagging behind [50, 51]. This is a case where regulation appears to have run ahead of scientific evidence, which in some cases is controversial [52].

All the above is for direct toxicity, which is the only way to assess the toxic potency of a chemical compound. Assessment of the toxicity of mixtures of chemicals started in the 1970s; this can be determined using the same protocols for acute and chronic toxicity [53] or using genomics [54]. Apart from determining the total toxicity of the mixture, it is important to assess whether the compounds show additive, synergistic or antagonistic toxicity, and if so, try to estimate the synergistic/antagonistic ratios [41, 55]. Additive toxicity occurs with compounds that have the same or similar mode of action, e.g. organophosphorus and carbamate insecticides [56], pyrethroids, neonicotinoids, triazines, etc. In such cases, a toxic equivalent (TE) with reference to a standard compound in the same group of chemicals is estimated [57]. TEs are then used for assessing the total toxicity of such mixtures in tissues, sediments or another matrix [58]. Among synergistic mixtures, the best and well known is that of piperonyl butoxide with pyrethroid insecticides, which occurs because of inhibition of the mono-oxygenase detoxification system in insects [59]. Recently, mixtures of ergosterol-inhibiting fungicides with cyano-substituted neonicotinoids have also shown enhanced synergism of these insecticides in honeybees [60].

Indirect impacts of chemical toxicity occur in nature [61]. Although it is often difficult to determine the causal relationship that exists between chemicals and the indirect effects they may have in ecosystems, well-designed microcosms and mesocosms, as well as field studies, have often proved such effects [62-64]. Indeed, ecological theory predicts that the elimination of predators inevitably results in a bloom of the primary prey species they control [65]. Equally, the elimination of key primary consumers leads to the uncontrolled growth of their feeding sources (e.g. weeds, algae), and this occurs when insecticides eliminate grazing planktonic species [66], but not necessarily to the demise of their predators, as the latter organisms can switch prey preferences [67]. Fungicides can eliminate fungal communities that have the important role of recycling organic wastes in nature and thus reduce the efficiency of the litter decomposition and mineralisation in soil [68]. Reduction of the insect food source after application of imidacloprid in agricultural environments over many years has led to the reduction of several bird species in the Netherlands [69, 70]. Even biological insecticides, such as Bt (*Bacillus thuringiensis*) applied to control mosquitoes, flies and other nuisance insects in wetland areas, inevitably reduce the bird populations that feed on those insects [71]. One of the practical applications of this theory in agriculture is the integrated pest management (IPM) [72]. Indeed, IPM is mindful of the disastrous consequences of pesticide application for pest control without consideration of the impacts on the whole ecosystem. Extreme examples are the massive explosion of brown plant hoppers (*Nilaparvata lugens*) in Indonesia during the 1970s following excessive application of insecticides that eliminated spiders and other predators of this rice pest [73] and the uncontrolled population growth of red-billed quelea (*Quelea quelea*) in areas of Botswana with high residue loads of fenthion, a persistent organophosphorus insecticide applied for pest control in agriculture [74]. Indirect effects, however, are not included in the first tier of regulatory risk assessment of chemicals; in some cases, they may be considered in the second tier (Fig. 1).

6. Exposure Assessment

The first step in the exposure assessment starts with characterisation of the chemical properties of a substance, as this will determine the main routes of exposure for organisms. Among the physical and chemical properties relevant for environmental risk are solubility in water and organic solvents; partitioning coefficients, in particular the octanol-water coefficient (K_{ow}); soil adsorption constants (e.g. K_{oc}); vapour pressure and volatilisation (Henry's law); dissociation constants (pK); and half-lives in several environmental matrices (air, water, soil, sediments and plant tissues) due to different processes: hydrolysis, photolysis or metabolism. In addition, mobility and leaching potential should be assessed, and this can be done by means of indices such as the ground ubiquity score (GUS), which combines soil adsorption properties and degradation in soil [75] (Table 1). The databases mentioned above (i.e. AGRITOX, Footprint and Pesticide Manual) are comprehensive sources of such data. Manufacturers of new products must provide this information – different from the material safety data sheets (MSDS) – to their regulatory authorities for risk assessment. Numerous laboratory measurements and tests must be carried out to obtain this kind of data, although in some cases, quantitative structure-property relationships (QSPRs) are used to derive approximate values when compounds have similar chemical structures.

Degradation pathways in animals are also important, as some chemicals are amenable to metabolic breakdown and produce metabolites, usually of less toxicity, that can be eliminated in urine and faeces. However, metabolites can also be as toxic as the parent compounds, in which case, they need to be assessed as independent toxicants. Often, a chemical may be degraded easily in water or plants but may be persistent in soil or *vice versa*. Recalcitrant chemicals are the ones that do not degrade readily in any of the environmental media, e.g. chlorinated compounds, including some neonicotinoid insecticides and herbicides. Persistent agrochemicals require special attention because their effects are not restricted to the time of their application to crops but also during a long period of time afterwards, which could be years. For example, in many developed countries, residues of DDE in soil are still causing egg-shell thinning in birds, even if the parent compound (DDT) has not been applied for more than 30 years [76].

Once the basic data are available, concentrations of the chemical in various matrices can be predicted after its application to a crop, using mathematical models developed for specific purposes [14]. Transport models can refer to a single medium (e.g. air) or to the fate and movement between media. Examples of the first type are the dispersion of particles and movement through air, which can be modelled by AgDRIFT [77] and Gaussian models [78], soil erosion and sedimentation of runoff [79] and leaching of water-soluble fractions into the soil profile, e.g. Pesticide Root-Zone Model (PRZM) [80] and others [81]. By contrast, multimedia models such as fugacity [82] aim at predicting the concentrations of chemicals in all environmental media. In any case, model predictions should be validated by comparing the outputs of the model to the actual measurements of residues in water, air, soil and organisms. However, for new chemicals this is not feasible, as residue amounts could be negligible or non-existent, so current regulatory risk assessment relies entirely on the predicted environmental concentrations (PECs) that result from such models. For old agrochemicals, on the other hand, measured residue levels would be preferable for two reasons: (i) a large body of actual residue data exists for different regions and times, as comprehensive monitoring surveys are carried out over many years and are available from the open literature, and (ii) relevant physicochemical and degradation data are often incomplete for many old compounds, thus hampering their environmental modelling. Each model output and survey data is considered a line of evidence for the risk assessment.

Essential information in regard to prediction of modelled data is the usage pattern of the chemical under consideration; this information is often not readily available, so it is often necessary to infer usage from sales figures. The total amounts of agrochemical applied determine ultimately the total load of residues to which organisms will be exposed. For persistent chemicals, residue loads will increase with time, so it is important to know the period of time they have been used. When all these factors are put together, the margins of error necessarily increase, and so variations of one order of magnitude in PECs are acceptable. The situation is no better for monitoring data, as the variability in residue loads between sites, regions, times of the year and among years can be even larger. Therefore, sensible exposure assessments should be based not so much on accurate concentration levels but on a range of concentrations for specific scenarios and situations. Statistical distributions of such residues help determine their probability of exposure in site-specific ERA [83].

PECs by themselves are not sufficient to estimate exposure within reasonable limits. The availability of chemicals is paramount, since only chemicals that are taken up by organisms can cause harm. Routes of exposure differ for aquatic and terrestrial organisms, and the estimated PECs should refer not only to the levels of chemicals in the environmental matrices but also the uptake by the organisms living there. In the aquatic environment, organisms are embedded in a matrix that contains pollutants, and their uptake through the gills and epidermis is almost constant. In this case, time of exposure really influences the chemical uptake, while toxicokinetics are influenced by surface area, porosity of the surface and volume/surface ratio. Also, the internal doses and effects with time can be estimated in accordance with Haber's rule (see above). Ingestion of residues in food appears to be a lesser route of exposure for zooplankton [84] and possibly other aquatic organisms. For terrestrial organisms, however, a more complex scenario is envisaged. Uptake of air pollutants may be constant through inhalation, but their residues in water and food sources (e.g. pollen, plant material, insects) are taken intermittently in small, discrete amounts and at different times. Dermal exposure, by contrast, is most relevant to animals living in the soil and sediment (e.g. earthworms, invertebrate larvae, etc.) or having unprotected skins (i.e. amphibians), whereas reptiles, birds and mammals would be somehow protected by the scales, feathers and fur that cover their bodies. Nevertheless, the lethal effects that follow deposition of pesticide spray droplets on birds suggest that concentrated hydrophobic compounds can enter easily their bodies through this route of exposure [85].

Table 1. Relevant physicochemical properties and environmental standards of 13 pesticides commonly found in Dutch surface waters.

Pesticide	Type[1]	Water solubility at 20 °C	Soil degradation, aerobic	Soil adsorption and mobility[3]		Leaching potential	Environmental quality standard (EQS) in the Netherlands[5]		
		(mg/L)	(DT50 in days)[2]	Koc (ml/g)	Kfoc (ml/g)	GUS index[4]	MTR (ng/L)	AA-EQS (ng/L)	Exceeded EQS > 5-fold in
Azoxystrobin	F	6.7	78	589	423	2.60	56		2012
Carbendazim	F	8.0	40	-	225	2.64		600	2012
Desethyl-terbuthylazine	m	327	70.5	-	78	3.90	2.4		2012
Dicamba	H	250,000	4.0	-	12.36	1.75	130		2010
Dichlorvos	I, A	18,000	2	50	-	0.69		0.6	2012
Dimethoate	I, A	39,800	2.6	-	28.3	1.06		70	2011
Esfenvalerate	I	0.001	44	5,300	-	0.45		0.1	2012
Fipronil	I	3.78	142	-	727	2.45	0.07		2012
Imidacloprid	I	610	191	-	225	3.76		67	2012
Metolachlor	H	530	90	120	163	3.49	200		2011
Permethrin	I	0.2	13	100,000	-	-1.11	0.2		2010
Pirimiphos-methyl	I, A	11	39	1,100	170	2.82		0.5	2010
Propoxur	I, A	1,800	79	30	-	4.79	10		2010

1. A = acaricide; F = fungicide; H = herbicide; I = insecticide; m = metabolite.

2. DT50 is the degradation time to 50 % of the initial amount.

3. K_{oc} and K_{foc} are organic-carbon sorption constants.

4. The GUS index (groundwater ubiquity score) is calculated as GUS = log DT50 x (4 – log K_{oc}). Chemicals with GUS > 2.8 are likely to leach, whereas if GUS < 1.8, leaching is unlikely.

5. The maximum tolerable risk (MTR) and annual average environmental quality standard (AA-EQS) are traditional Dutch water quality standards.

One way of finding out the availability of pollutants to organisms is by using biomarkers of exposure. Many biomarkers have been developed in the last two decades with the aim of measuring the extent of chemical exposure in organisms [86]. Usually they measure the physiological response to chemicals with the same mode of action, i.e. the cholinesterase assay determines the proportion of acetyl-cholinesterase enzyme bound to cholinesterase inhibitor insecticides [87]. Biomarkers based on inhibition of the detoxification mechanisms (P450 mono-oxygenases, glutamate-S-transferases, etc.) are less specific, as many different compounds can trigger the same response. Unless the chemical source is known, biomarkers can only provide evidence of the existence of a group of contaminants that are taken up by the organisms [88, 89]. For this reason, they are useful to prove the availability of such chemicals, not for the identification of the individual compounds.

7. Risk Assessment: First Tier

Having obtained the above information, the chemical assessor is now in a position to evaluate the possible impacts of a chemical in the environment. The first hurdle assessors encounter is how to put together the disparate data gathered through the various lines of evidence from the toxicity and exposure assessments.

Whatever methods are used to evaluate the possible risks of a chemical, assessors must consider a range of scenarios: from a typical or most likely PEC under standard agricultural practices to a worst-case scenario that considers the highest possible PEC to be found in a particular environmental medium or under special conditions. Modelled PECs as well as monitoring residue data can be used for this first-tier assessment (Table 1).

For a start, different LC50 and LD50 values are derived for species belonging to disparate taxa, with a range of variation spanning several orders of magnitude! So, which dataset should be used in the risk assessment? One way of solving this problem is to evaluate representative species of each environmental media, and this has traditionally been done with the hazard quotient (HQ). OECD hazard assessments require at least one species from three different taxonomic groups (e.g. alga, crustacean, fish) for this first tier. Thus, for each environmental medium, one to three HQ values may help define the range of negative impacts. These HQ values should be derived for different exposure routes in the case of terrestrial organisms (e.g. dietary, inhalation or contact by dermal exposure), whereas a single value may be sufficient for aquatic organisms. Two shortcomings are found with this approach. Firstly, it is simplistic and often unrealistic, given that PECs for a range of scenarios typically vary up to one order of magnitude, and this variability can introduce a swing from being "safe" to "unsafe" if the resulting HQ values change from <0.1 to 0.1–1.0. Thus, there is an inherent large uncertainty in this approach even if the toxicity endpoints used are consistent and accurate. Secondly, a given chemical may be considered safe for the representative organism on which the toxicity data are based, but not safe for other organisms living in the same environment. An example of this is imidacloprid, which renders values of HQ below 0.0001 for standard *Daphnia* spp. at relevant concentrations of 5 μg/L in rice fields but has HQ values of 0.01 for epibenthic ostracods and benthic chironomid larvae [90] and values of 0.2–7.0 for larvae of mayflies [91]. This highlights the problem of using data for standard test species in the first tier to make decisions regarding the registration of a chemical; and yet most regulatory assessments still use this method.

A second and more plausible approach is to use all the toxicity data available for a chemical at once, by constructing an SSD – risk assessments using this approach are termed probabilistic risk assessments (PRA). For agrochemicals, SSDs are usually built using acute LC50 or LD50 data. Often an assessment factor is applied to give a chronic equivalent when the NOEL values are not available. In such cases, a factor of 10 or a derived acute-to-chronic ratio is commonly used. The SSD shows the overall range of toxicities for all species tested in the same environmental medium (e.g. aquatic; Fig. 2), so the assessor can determine the levels of chemical that would not have a serious impact on the majority of species. An advantage of the SSD is that it allows estimation of the number of species within a particular medium that would be seriously affected for any residue concentration of a given chemical. Since the latter residue levels can vary spatially, seasonally and with the passing of time, this relationship between residues in the environment and species affected (which includes their identity as well) can be very useful for risk management in site-specific ERA. In regulatory assessments, the threshold concentration hazardous to 5 % of species (HC5) is defined and compared to the PECs in that particular medium [92]. The term protective concentration for 95 % of species (PC95), which equates to HC5, is also used. Nevertheless, the probabilistic approach is rarely used for registration of new agrochemicals, since the range of toxicity data available for such compounds is usually insufficient to generate a valid SSD. They are mostly used for site-specific risk assessments, in which case, actual residue data from monitoring at the site or region are also used to complement the predicted exposures. PRAs are available for residues of the herbicide atrazine in waters [93] and for insecticides such as chlorpyrifos [94], pyrethroids [95] and endosulfan [96] and other agricultural chemicals [97, 98].

The probabilistic approach is not exempt from criticism, as the protective thresholds set are subjective and have no ecological basis. It could well be that the most important species to be protected are the most sensitive, as it happens with pollinators and parasitic wasps exposed to neonicotinoids and other compounds [99]. One could argue that setting a threshold that protects 100 % of species is the most adequate solution to this problem, but it is indefensible to estimate such limits when the datasets are often incomplete, as it occurs with most chemicals, and also because the statistical error increases at the bottom of the SSD curve. Obviously, the accuracy of the SSD relies entirely on the number of species tested and the taxa covered in the dataset. Another criticism is that the actual concentrations measured in surface waters are typically underestimated because the peak concentrations of pesticides, which have the highest impact, are often missed, thus casting doubts about the reliability of the exposure predictions on which the probabilistic approach is based [100]. One solution is to build a cumulative graph of either measured or predicted environmental concentrations for different exposure scenarios, thus allowing estimation of probabilities of exposure in a similar way as the SSD [15]. This approach has successfully been used in ERA of residual herbicides that have a long history of usage [83, 101], for which sufficient residue data are available. Another solution to this problem is the use of passive samplers that produce an integrated measurement of residues over a period of time [102]. Equally, passive samplers are also used for estimation of residue concentrations in the air phase [103]. In any case, SSDs provide a greater degree of accuracy than HQs in determining safe levels for chemicals in the environment. In fact, this approach has been used to set water quality guidelines in many countries, although it is more commonly used in the second tier of assessment [98].

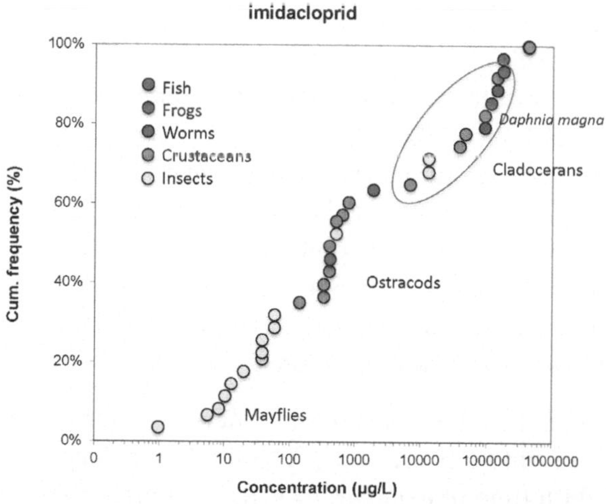

Figure 2. Species sensitivity distribution (SSD) of the insecticide imidacloprid for aquatic species (data source: ECOTOX database).

A third approach is to use risk indices that combine toxicity values of a chemical to several organisms, its persistence in the environment, the frequency of detection of its residues, the likelihood of drift and other characteristics that are deemed important. Indices of risk were first introduced by Metcalf to rank the risk of pesticides to fish, birds, mammals and bees [104]. They are useful for comparing the potential risks of several compounds applied to the same environment, so that farmers, pest control operators and other users of pesticides may choose chemicals that have the least risk to the environment. Examples of indices are the Pesticide Impact Rating Index (PIRI), used by managers and pesticide users as a simple guide for selecting agricultural chemicals that are less damaging to the environment [105]; the ecological relative risk (EcoRR), designed to discriminate between chemicals that may have ecological impacts at a site-specific environment [106]; other risk indices, along with pesticide use information, were reasonable predictors of bee poisoning incidents compiled over a 21-year period in the United Kingdom [107]. Many indices have been proposed over the years (see review by Levitan et al. [108]), and some of them have been used in regulatory risk assessments in European countries [109]. It is important to realise that such indices are only indicative of relative risks among compounds and cannot predict the actual risk that a particular agrochemical may have in a given environment.

7.1. New Approaches to Risk Assessment

First-tier assessments of agrochemicals are mostly based on acute toxicity data and a range of concentrations for different scenarios. Rarely do they consider chronic toxicity; this is most commonly estimated for mammals. In fact, survival over time is not considered for the vast majority of organisms tested. Indeed, one aspect that none of the methods used, whether the HQ, the SSD-probabilistic method or indices, have considered is toxicity in time. This may be due to the assumption that chemicals only cause toxicity on a dose-dependent manner, while the ratio between acute and chronic LC50s or LD50s is assumed to be fixed [110]. Recently, however, it was demonstrated that such ratios vary depending on the models used to estimate chronic toxicity [111]. Moreover, it is known that certain chemicals exhibit time-cumulative effects with time, such that infinitesimal concentrations can cause toxicity upon chronic exposure [40]. For these chemicals, which can be identified by a high acute-to-chronic ratio [91], the current methods of risk assessment fail to produce a plausible and realistic evaluation of their risks to the environment.

A new approach has been proposed, therefore, for assessing the risk of chemicals with time-cumulative toxicity. The method is based on the relationship of LC50 or LD50 with time of exposure, which is defined mathematically by a log-to-log regression between the dose and the time to reach 50 % effect (ET50) [112],

$$L_n ET50 = a + b Ln LD50 \quad (1)$$

The inverse of the slope of that regression (b) is the power exponent of the Druckrey-Küpfmüller equation [113, 114]

$$C * tn = constant \quad (2)$$

where n =1/b of the regression (1) and C is the PEC or measured concentrations of the chemical in the environment. Since the exponent n for a chemical can be determined using time-to-event bioassay data, the times to reach 50 % effect can be estimated for any given concentration in the media or dose ingested by organisms.

In fact, the slope in equation (1) helps determine whether a chemical may have time-cumulative toxicity or not, because only the chemicals with values of n >1 are included in this special category [40], thus making the task of the assessor easier.

This method, therefore, estimates the time to reach a critical endpoint (e.g. 50 % mortality) based on the PECs or measured concentrations found in the environmental media. It has been applied to evaluate the risk of two neonicotinoids (imidacloprid and thiamethoxam) to honeybees and bumble bees [115]. The risks of these insecticides to bees, as evaluated by this new approach, are in better agreement with the observed declines of these insects in countries where neonicotinoids are used than the risks evaluated by classical methods. As indicated above, only the chemicals showing time-cumulative toxicity need to be considered here; for most compounds, the current probabilistic methods are appropriate.

Another aspect that has been neglected for too long is the inclusion of toxic metabolites and transformation products in risk assessments [116]. Certainly, a number of toxic pollutants, including insecticides (e.g. aldicarb, dimethoate, endosulfan, fipronil, heptachlor, imidacloprid, thiamethoxam and others), persistent pollutants such as DDT and a range of industrial chemicals, can be converted to degradates that are as toxic as the parent compounds or even more. For example, organophosphorus pesticides that contain sulphur double bonds can be more toxic when oxygen replaces the sulphur during degradation; such is the case of chlorpyrifos, profenofos and parathion. A proper risk assessment of these chemicals should include not only the original compound but the total toxic burden, i.e. parent + toxic metabolites [117]. While this area of assessment is still incipient in risk evaluations of agrochemicals, similar procedures as those used with mixtures should be devised to assess their toxicity.

8. Risk Assessment: Second Tier

It is envisaged that not all chemicals will pass the first tier without problems. Concerns inevitably arise from the lack of sufficient data regarding toxicity to some species that may be deemed important in a particular environment or from the lack of chemical data, which results in poor predictability of the fate of a new chemical. Too many uncertainties in the toxicity and exposure steps result in unrealistic risk evaluations. When that happens, regulators may request that manufacturers provide further evidence on whether a chemical may or may not have negative impacts on the environment.

The purpose of the second tier is, therefore, to obtain more data about the behaviour of the chemical so as to fill gaps in knowledge and resolve the uncertainties and consequences of its release in the environment. Different methods may be used to achieve this, each of them seeking evaluation in a realistic scenario:

- Field trials

- Mesocosms and microcosms
- Sublethal effects and chronic studies
- Side effects – endocrine disruption
- Life-cycle effects
- Population modelling

8.1. Field trials

Field trials, using typical and worst-case concentrations of the chemical, are useful to measure realistic concentrations and dissipation rates in water, soil, air and the movement between environmental compartments, e.g. leaching through the soil profile [118], volatilisation [119], chemical loads in runoff [120], eroded materials and sediments [121] and others. For concentrations in water and air, passive samplers give better estimates than individual grab samples, as the samplers integrate peaks from pulses as well as lower concentrations over time. Indeed, concentrations of pesticides in water measured using passive samplers are better correlated with the effects observed in aquatic organisms [122]. Air passive samplers are useful to study the transport of volatile chemicals such as endosulfan and persistent organochlorines [103]. A drawback of field trials is that factors other than the concentration of the agrochemical applied cannot be controlled. This introduces uncertainty about the effects on organisms, which could be due to the chemical applied or to other environmental factors, whether in combination or alone. In addition to this problem, it is almost impossible to monitor all individuals present in a field and its surroundings; consequently, the effects on organisms are underestimated. This may explain the apparent lack of mortality in birds in field trials where organophosphorus and other insecticides are applied to agricultural fields, since most of the fatalities go undetected [123]. Equally, several field trials with systemic insecticides appeared not to cause much impact in honeybee colonies placed in the surrounding of treated fields [124-126], but this is partly because bees that died or got lost in the field – due to disorientation – could not be counted [127], while at the same time, the hives keep producing more worker foragers to compensate for those losses. In this case, only by tagging of individual bees with special tracking devices may render accurate estimation of losses and foraging activity [128, 129]. As it turns out, bee losses are larger than originally expected from former field trials with those insecticides [130, 131]. Another problem with field studies of bees is that often the controls used are contaminated with a diverse array of pesticides [132], so the effects of the chemical under study cannot be distinguished from the controls or are confounded by other pesticide contaminants.

Field studies can include surveys where chemical and biotic data are collected at the same place and time. Such data can be analysed to determine whether or not the chemical pollutants found in a particular location have reduced its biodiversity. That requires, of course, a reference to natural or near-natural conditions. For instance, in order to determine the extent of pollution in water bodies, the European Water Framework Directive requires such definition before the extent of contamination by stress gradients (chemical or otherwise) can be established. To this purpose, the Species at Risk (SPEAR) index was developed to express the risk of pollutants to aquatic ecosystems. It considers the overall toxicity of pollutants measured in terms of toxic units with reference to *Daphnia magna* ($TU_{Daphnia}$) and has a built-in distinction between sensitive and non-sensitive species. SPEAR can be used as an indicator of the pollution status of rivers or other aquatic ecosystems [133].

8.2. Mesocosms and microcosms

Mesocosms and microcosms are model ecosystems but on a small scale. Given that most factors are controlled by the researchers, the effects of chemicals on organisms can be determined with more certainty than in field trials. Also, organisms are usually confined to the small experimental area (a few square metres per unit) and thus can be counted more reliably. Indirect effects on other species and communities are usually detected in this kind of experiments [134, 135], whereas such effects would be difficult to observe and measure in field trials. One positive feature of mesocosms and microcosms is that they avoid contamination of the environment, something that field trials cannot do. The experimental units are self-contained, and contaminants in the water are eliminated from any runoff exiting the system or they are recovered. Another advantage of these systems is that can be replicated using practically the same conditions, whether outdoor or indoors, whereas field trials are subject to the inevitable variability existing among fields. In both, field trials and mesocosms, the use of controls without chemical treatment is imperative as a reference.

The data generated in these systems usually include dissipation half-lives of the chemicals in various matrices, changes in population and biodiversity in the communities under monitoring, and occasionally, some functional traits of the ecosystem (e.g. productivity) can also be obtained [64]. Data on individual species can be analysed using traditional methods to estimate the median effect concentrations in the field, and once these are obtained, an SSD can be built to determine the HC5 for the communities. Following this procedure, the laboratory toxicity data can then be compared to the field or mesocosms toxicity data to either prove or correct the findings [136]. Effects of chemicals on communities can be measured in a similar way as in typical toxicity tests, and dose-response relationships can be obtained for some endpoints, typically the abundance or species richness [137]. The communities' data can also be subjected to multivariate statistical analysis such as principal component analysis (PCA) or, even better, principal response curves (PRCs) [138]. The latter analytical tool helps determine the extent of the changes occurring in the community under exposure to the chemical(s) as well as their statistical significance. PRCs have been used to study the recovery of populations in agricultural ditches exposed to pesticides [139] and to compare the ecological effects of different insecticides applied to rice paddies [140].

Mesocosms have been instrumental in detecting effects of chemicals in populations and communities of macro-invertebrates [141], which otherwise could not be predicted using standard toxicity tests in the laboratory. This is important, since the functionality of an ecosystem depends on the structure and relationships between all the species that compose it. Even if the functionality of an ecosystem can be restored after the chemical is withdrawn, often the structure is changed, so the recovery of populations does not restore the disturbed community to the same structure it had before. This is because some species have been more affected than others or have been eliminated and need to be replaced by other species. This can be demonstrated by using similarity indices, a tool commonly used in ecological studies that can also be applied to assess the changes occurring in a community of arthropods during and after the exposure to insecticides [142]. Biodiversity indices, on the other hand, are poor indicators of ecosystem changes [137, 143]. Similarity indices have also been used to estimate chemical thresholds that protect ecological communities from contaminants in freshwater systems [144]. Structural changes are usually detected by such similarity indices or by measuring the species richness, both of which are good indicators of ecosystem health and can be used to assess the impacts of pollutants on ecosystems [145]. However, among the difficulties and drawbacks of mesocosms can be cited their high cost and need for more replication, especially under different climatic or environmental conditions. This is because sometimes replicates produce variable results (e.g. communities go in different directions) that defy statistical analysis.

8.3. Sublethal effects and chronic studies

Laboratory experiments using sublethal doses/concentrations of chemicals are used to determine a variety of sublethal effects that were not considered in the first tier, which was only concerned with survival under exposure to lethal doses of chemicals. Sublethal effects can be dose dependent if they are related to the mode of action of the compound, e.g. behavioural disturbances caused by neurotoxic insecticides [146-148]. In such cases, sublethal effects can be assessed using HQ or other standard numerical methods used in tier I. However, sublethal effects often resist an explanation based on the known mode of action of the chemical concerned. For instance, is the anti-feedant behaviour [149] of imidacloprid in arthropods the result of the agonistic action of this insecticide on the nicotinic receptors? If so, it might be possible to establish a dose-response relationship for low exposure doses, but if not, how can the risk of sublethal doses of this insecticide be assessed in a quantitative manner? To date, no established method of numerical analysis of sublethal effects exists in the public literature, in spite of its importance for the long-term viability of natural populations of insects [150] and other organisms. In any case, sublethal effects are difficult, if not impossible, to assess in field trials and mesocosms. There are many examples in the published literature of sublethal effects of man-made chemicals, and they are usually reported in extensive reviews carried out by either independent or regulatory authorities. However, unlike lethal toxicity data, sublethal data does not produce robust statistical estimates of effects. Also, this kind of evidence is not always available for newly developed agrochemicals – information on sublethal effects comes usually a long time after a product has been approved by the authorities and used extensively. More often than not, research on sublethal effects is prompted by a series of negative impacts observed with particular products after many years: the reproductive impairment in birds caused by DDE (a metabolite of DDT) took about 20 years to be elucidated [151]; and the current demise of insect populations and communities as a result of the widespread use of neonicotinoids has taken some ten years to be understood [152].

8.4. Side effects – Endocrine disruption

Endocrine disruption is usually detected in experimental work of the second tier. This is an aspect that has been missing in ERAs until recently. One feature of EDC effects is that the potency of individual chemicals may be low but the effect can exceed thresholds with mixtures of additive EDC potency. However, endocrine disruption is for the most part a side effect unrelated to the main mode of action of the chemical concerned, and when present, its effects appear to follow a typical monotonic, dose-response relationship like any other mode of action. Some authors advocate that the mode of action of EDCs is non-monotonic but rather follows a U-shape curve similar to the response of natural hormones and micronutrients [153-155]. This unusual relationship has been disputed by many authorities in toxicology and ecotoxicology [52, 156], who argue that EDCs should be treated in the same way as any other toxic chemical, i.e. following the same evaluation methods for their risk as outlined above, using either HQs, SSDs or time-cumulative methods.

8.5. Life-cycle effects

As a corollary of the previous sublethal and endocrine studies, life-cycle parameters can be determined to assess the impact of a chemical through the various stages of development of an organism [157]. Effects on survival, reproduction and development are thus integrated on a life table. This is particularly relevant for invertebrates, which often change from larvae to adults by experiencing several moulting stages or through metamorphosis.

8.6. Population modelling

Population modelling helps estimate the long-term consequences of regular, chronic chemical impacts on populations of organisms. In essence, data obtained at the individual level need to be extrapolated for effects at the population level using life table evaluations [158, 159]. Although more theoretical than the previous methods, they are by no means less realistic, and they can be validated with experimental work appropriately designed for that purpose. Common population endpoints used are abundance, population growth rate [160, 161] and the chance of population extinction [162]. Two types of models have been suggested: simple life-history models distinguishing between life-history stages of juveniles and adults [163] and spatially explicit individual-based landscape models [164]. Data can be gathered from experimental mesocosms or from historical records. For example, a demonstration that the decline in populations of sparrowhawks (*Accipiter nisus*) and kestrels (*Falco tinnunculus*) in Britain between 1963 and 1986 was due to dieldrin contamination was done using growth rates of those birds and analysing their life-history data on body residues [165]. The performance and long-term survival of honey bee colonies have also been modelled after exposure to systemic insecticides [166], as well as the breeding success of wood mice and skylarks exposed to a fictitious fungicide [167].

8.7. Evaluation process in the second tier

Unlike the straightforward mathematical methods used to analyse information for the first tier of the risk assessment, there aren't any mathematical procedures to deal with the risks derived from sublethal effects, indirect effects on communities, endocrine disruption, etc. Consequently, all of the evidence gathered for this second tier must be evaluated using the logical methods commonly used in the WOE approach. The exception, perhaps, is the PERPEST model [168], which was developed to predict risk of pesticides in freshwater ecosystems. It simultaneously predicts the effects of a particular concentration of a pesticide on various community endpoints. The model relies on case-based reasoning, a technique that solves new problems by using past experience (e.g. published microcosm or mesocosm experiments). In order for the model to do that, empirical data extracted from the literature are stored in a database of freshwater ecotoxicity studies, which are updated regularly.

Apart from PERPEST, whenever there is some evidence of negative effects of a chemical on a species or an ecosystem, risk assessors must weigh that evidence against the benefits that the chemical may have for human life, the general environment and agriculture – there may be some negative effects to certain organisms or certain areas, but not necessarily to entire ecosystems. How serious are the effects and how widespread their threats are questions for the assessors to ask and evaluate using their best professional judgement. Here, the purpose of the risk assessment influences the assessor's evaluation. For example, in regulatory assessments, the aim is to minimise the impact of a particular chemical on ecosystems to levels accepted by the community, which is not necessarily the same as protecting the integrity of the ecosystems. Thus, some chemicals can be tolerated if the benefits they have for our lifestyle offset their impacts to specific environments, i.e. agricultural areas that are sacrificed for the "common good". The task of assessing a chemical's risk in this context is not easy, as the value of the environment cannot be measured directly in terms of money [169], whereas the benefits of agricultural and industrial production are more tangible. Not surprisingly, there are many discrepancies between the regulations of various countries, even for the same agricultural chemical [170] because they are based on different human judgements. In practice, many hazardous agrochemicals are registered as long as certain precautions and management options are put in place. This similarly occurs with pharmaceuticals and industrial chemicals.

Other important questions to be asked during the evaluation process are how much of the chemical will be used and how often, since the concentrations or doses in the environment will depend entirely on the application rate and frequency of usage. Residue loads of persistent agrochemicals may accumulate in soil over the years until their concentrations or doses may reach sublethal or lethal levels to certain organisms and, in the case of herbicides, some agricultural crops. In such cases, regulators may suggest to restrict the application of such products to avoid their accumulation in the environment.

A final question, often raised by risk assessors of agrochemicals, refers to the different behaviours that a given compound may have in regions of the world that vary in climatic and other environmental conditions. Should the special conditions of those environments be taken into consideration at the time of allowing a pesticide to be used in a country? Currently, the bulk of insecticides are being used in developing countries, most of which are located in the tropical or subtropical regions of the world, where warm and humid conditions may affect their dissipation and exposure to organisms. Would the risk of such chemicals in tropical countries be lesser than their risk in temperate and cold countries? It appears that most chemicals pose a similar risk in either region because the increasing losses by microbial degradation or volatilisation are usually counterbalanced by greater desorption and movement of residues into the aquatic environment [171]. Using tropical taxa for toxicity testing in those countries has also been suggested, arguing that tropical species may differ in susceptibility to pesticides. However, most tropical species do not differ in sensitivity from their temperate counterparts in regard to agrochemicals [163], unlike what happens with metal contaminants [172, 173]. Overall, there is no convincing evidence that, in regard to environmental impacts of pesticides, agrochemical products used in tropical countries should be evaluated differently from any other country.

9. Risk Management

An important issue that stems from the evaluation of risks of agrochemicals is the management options available to mitigate or eliminate the hazards that these products introduce into the environment. Chemical companies should be responsible for establishing safety guidelines that ensure a continuous usage of their products does not harm the environment. These guidelines should be reviewed by the regulatory authorities together with all other relevant information for the risk evaluation of chemicals.

The most basic management option for any agrochemical is the application rate and frequency of usage of a product. Current pesticide labels specify the highest application rates to be used on specific crops, but these are based on the efficacy of the chemical to protect such crops against insect pests, diseases or competition from weeds, rather than on environmental grounds. One can assume that if the recommended rates are approved by regulatory authorities, they do not pose a serious harm to the environment; but is this always true? For newly developed chemicals, the manufacturing company must prove this is the case. However, for most chemicals already in the market, an ERA should be carried out with the aim of proving the safety of such rates. When the environment is compromised, restrictions to the frequency of application should be established so as to ensure a product is not applied *ad libitum*.

Other management options, often considered in IPM programs, may include i) restrictions to the type of crops the chemical products could be applied; ii) limits to the frequency of application of a product during a cropping season; iii) specific instructions about the mode of application, e.g. type of spray nozzles, planters, etc.; iv) timing of applications to avoid harm to bees; v) establishment of vegetated barriers or other types of buffer zones to capture spray drift that could fall onto neighbouring land or water bodies; and vi) establishment of vegetated ponds and/or sediment traps to capture runoff loaded with residues and allow the natural remediation of these [174]. There is ample evidence that some of these measures have proven very effective in reducing the impacts of pesticides in the environment [175, 176].

Finally, most product labels include some basic management practices. However, in many developing countries, these instructions are ignored because they are written in foreign languages or, more commonly, due to the illiteracy of farmers and applicators [177]. Label instructions are, therefore, insufficient to ensure a proper management of toxic chemicals such as pesticides.

10. Conclusions

Determining the risk of chemicals in the environment requires a complex evaluation of many different lines of evidence. We have examined here the processes leading to such an evaluation and criticised the main deficiencies of the current systems. Bearing in mind that no system is perfect, as the risk process is essentially a human evaluation, we have also pointed out new approaches that could help improve chemical assessments in the future.

One area that will always be crucial to any chemical risk assessment is the accuracy and reliability of chemical and toxicity data. Without proper data, the evaluations would fail to determine realistic risks under possible scenarios. In this context, modelled predictions of exposure must be validated with factual data, whatever the conditions. A second area for improvement is the science underpinning the toxicity of chemicals. Toxicology as a science related to human health has a long history, but its application to environmental science lags behind. Part of the problem is that ecotoxicology relies too much on acute toxicity data obtained under laboratory conditions. In the absence of any other information, such data are used to evaluate risks in the first tier of the risk assessment, so mistakes are commonly made. While this shortcoming has been recognised a long time ago, the inclusion of new risk approaches that take into consideration long-term, sublethal, endocrine and population effects is met with resistance, perhaps because the new information thus gathered cannot be easily analysed using mathematical models. Time-dependent toxicity has made great advances in recent decades, and yet its incorporation in risk assessment is lagging. These aspects need more attention by risk assessors and regulators of new agrochemicals if risk assessments are to be more realistic.

References

1. Sánchez-Bayo F. Sources and toxicity of pollutants. In: Sánchez-Bayo F, Van den Brink PJ, Mann R, editors. Ecological Impacts of Toxic Chemicals. Online: Bentham Science Publishers; 2011. p. 3-12.
2. Chapman PM. Determining when contamination is pollution - Weight of evidence determinations for sediments and effluents. Environment International. 2007;33(4):492-501.
3. Sánchez-Bayo F. Impacts of agricultural pesticides on terrestrial ecosystems. In: Sánchez-Bayo F, Van den Brink PJ, Mann R, editors. Ecological Impacts of Toxic Chemicals. Online: Bentham Science Publishers; 2011. p. 63-87.
4. Mineau P, Whiteside M. Pesticide acute toxicity is a better correlate of U.S. grassland bird declines than agricultural intensification. PLoS One. 2013;8(2):e57457.
5. Suter-II GW, editor. Ecological Risk Assessment. Chelsea, MI: Lewis Publishers; 1993.
6. Chapman JC. The role of ecotoxicity testing in assessing water quality. Australian Journal of Ecology. 1995;20:20-7.
7. Hoeven NVd. How to measure no effect. Part III: statistical aspects of NOEC, ECx and NEC estimates. Environmetrics. 1997;8:255-61.
8. Kooijman SALM, Bedaux JJM, Slob W. No-effect concentrations as a basis for ecological risk assessment. Risk Analysis. 1996;16(4):445-7.
9. Urban DJ, Cook NJ. Ecological risk assessment. Standard evaluation procedure of the Hazard Evaluation Division. Washington D.C.: Office of Pesticide Programs, Environment Protection Agency, 1986.
10. Lee-Steere C. Environmental Risk Assessment Guidance Manual for Agricultural and Veterinary Chemicals. Australian Environment Agency Pty Ltd, 2009.
11. Guidelines for testing of chemicals - Daphnia sp., Part I: Acute Immobilisation Test, 202. Sect. 2. Effects on Biotic Systems (1984).

12. OECD Guidelines for testing of chemicals - Daphnia magna Reproduction Test, 211 (1998).
13. Escher BI, Ashauer R, Dyer S, Hermens JLM, Lee J-H, Leslie HA, et al. Crucial role of mechanisms and modes of toxic action for understanding tissue residue toxicity and internal effect concentrations of organic chemicals. Integrated Environmental Assessment and Management. 2011;7(1):28-49.
14. Meent Dvd, Hollander A, Peijnenburg W, Breure T. Fate and transport of contaminants. In: Sánchez-Bayo F, Van den Brink PJ, Mann R, editors. Ecological Impacts of Toxic Chemicals. Onine: Bentham Science Publishers; 2011. p. 13-42.
15. Solomon K, Giesy J, Jones P. Probabilistic risk assessment of agrochemicals in the environment. Crop Protection. 2000;19(8-10):649-55.
16. Aldenberg T, Jaworska JS. Uncertainty of the hazardous concentration and fraction affected for normal species sensitivity distributions. Ecotoxicology and Environmental Safety. 2000;46(1):1-18.
17. EPA U. Guidelines for carcinogenic risk assessment. 1986. p. 38.
18. EPA U. Guidelines for ecological risk assessment. 1998. p. 188.
19. Good IJ. Weight of evidence and the Bayesian likelihood ratio. In: Aitken CGG, Stoney D, editors. The Use of Statistics in Forensic Science: CRC Press; 1991. p. 85-106.
20. Weed DL. Weight of Evidence: A review of concept and methods. Risk Analysis. 2005;25(6):1545-57.
21. Linkov I, Loney D, Cormier S, Satterstrom FK, Bridges T. Weight-of-evidence evaluation in environmental assessment: Review of qualitative and quantitative approaches. Science of the Total Environment. 2009;407(19):5199-205.
22. Chapman PM. Traditional ecological knowledge (TEK) and scientific weight of evidence determination. Marine Pollution Bulletin. 2007;54:1839-40.
23. Landis WG, Chapman PM. Well past time to stop using NOELs and LOELs. Integrated Environmental Assessment and Management. 2011;7(4):vi-viii.
24. Jager T. Bad habits die hard: The NOEC's persistence reflects poorly on ecotoxicology. Environmental Toxicology and Chemistry. 2012;31(2):228-9.
25. Kooijman SALM. An alternative to NOEC exists, but the standard model has to be abandoned first. Oikos. 1996;72(2):310-6.
26. Fox DR. A Bayesian approach for determining the no effect concentration and hazardous concentration in ecotoxicology. Ecotoxicology and Environmental Safety. 2010;73:123-31.
27. Zwart DD, Slooff W. The Microtox as an alternative assay in the acute toxicity assessment of water pollutants. Aquatic Toxicology. 1984;4(2):129-38.
28. Calleja MC, Persoone G, Geladi P. Comparative acute toxicity of the first 50 multicentre evaluation of in vitro cytotoxicity chemicals to aquatic non-vertebrates. Archives of Environmental Contamination and Toxicology. 1994;26:69-78.

29. Boobis AR, Daston GP, Preston RJ, Oli SS. Application of key events analysis to chemical carcinogens and noncarcinogen. Critical Reviews in Food Science and Nutrition. 2009;49(8):690-707.
30. Breteler RJ, Williams JW, Buhl RL. Measurement of chronic toxicity using the opossum shrimp Mysidopsis bahia. Hydrobiologia. 1982;93(1-2):189-94.
31. Sinha N, Lal B, Singh TP. Pesticides induced changes in circulating thyroid hormones in the freshwater catfish Clarias batrachus. Comparative Biochemistry and Physiology, Part C. 1991;100(1-2):107-10.
32. Colborn T. A case for revisiting the safety of pesticides: A closer look at neurodevelopment. Environmental Health Perspectives. 2006;114(1):10-7.
33. Tomlin CDS. The e-Pesticide Manual. Surrey, U.K.: British Crop Protection Council; 2009.
34. Baas J, Jager T, Kooijman B. Understanding toxicity as processes in time. Science of the Total Environment. 2010;408(18):3735-9.
35. Kooijman SALM, Bedaux JJM. The Analysis of Aquatic Toxicity Data. Amsterdam, The Netherlands: VU University Press; 1996.
36. Bliss CI. The relation between exposure time, concentration and toxicity in experiments on insecticides. Annals of the Entomological Society of America. 1940;33(4):721-66.
37. Newman MC, McCloskey JT. Time-to-event analyses of ecotoxicity data. Ecotoxicology. 1996;5(3):187-96.
38. Mackay D, Puig H, McCarty LS. An equation describing the time course and variability in uptake and toxicity of narcotic chemicals to fish. Environmental Toxicology and Chemistry. 1992;11(7):941-51.
39. Ashauer R, Escher BI. Advantages of toxicokinetic and toxicodynamic modelling in aquatic ecotoxicology and risk assessment. Journal of Environmental Monitoring. 2010;12(11):2056-61.
40. Tennekes HA, Sánchez-Bayo F. The molecular basis of simple relationships between exposure concentration and toxic effects with time. Toxicology. 2013;309:39-51.
41. Escher BI, Hermens JLM. Modes of action in ecotoxicology: their role in body burdens, species sensitivity, QSARs, and mixture effects. Environmental Science & Technology. 2002;36(20):4201-17.
42. Jager T, Albert C, Preuss TG, Ashauer R. General Unified Threshold Model of Survival - A toxicokinetic-toxicodynamic framework for ecotoxicology. Environmental Science & Technology. 2011;45:2529-40.
43. Ashauer R, Hintermeister A, Caravatti I, Kretschmann A, Escher BI. Toxicokinetic and toxicodynamic modeling explains carry-over toxicity from exposure to diazinon by slow organism recovery. Environmental Science & Technology. 2010;44(10):3963-71.

44. Tennekes HA, Sánchez-Bayo F. Time-dependent toxicity of neonicotinoids and other toxicants: Implications for a new approach to risk assessment. Journal of Environmental & Analytical Toxicology. 2012;S4:S4-001.
45. Keith LH. Environmental Endocrine Disruptors. New York: John Wiley & Sons, Inc.; 1997. 1232 p.
46. Guillette LJ, Gross TS, Masson GR, Matter HM, Percival HF, Woodward AR. Developmental abnormalities of the gonad and abnormal sex hormone concentrations in juvenile alligators from contaminated and control lakes in Florida. Environmental Health Perspectives. 1994;102(8):680-8.
47. Stoker C, Zayas MA, Ferreira MA, Durando M, Galoppo GH, Rodríguez HA, et al. The eggshell features and clutch viability of the broad-snouted caiman (Caiman latirostris) are associated with the egg burden of organochlorine compounds. Ecotoxicology and Environmental Safety. 2013;98:191-5.
48. Frische T, Bachmann J, Frein D, Juffernholz T, Kehrer A, Klein A, et al. Identification, assessment and management of "endocrine disruptors" in wildlife in the EU substance legislation - Discussion paper from the German Federal Environment Agency (UBA). Toxicology Letters. 2013;223(3):306-9.
49. Colborn T, Vom Saal FS, Soto AM. Developmental effects of endocrine-disrupting chemicals in wildlife and humans. Environmental Health Perspectives. 1993;101(5):378-84.
50. Lewis RW. Risk assessment of "endocrine substances": Guidance on identifying endocrine disruptors. Toxicology Letters. 2013;223(3):287-90.
51. Kavlock RJ, Daston GP, DeRosa C, Fenner-Crisp P, Gray LE, Kaattari S, et al. Research needs for the risk assessment of health and environmental effects of endocrine disruptors: A report of the U.S. EPA-sponsored workshop. Environmental Health Perspectives. 1996;104(Suppl. 4):715-40.
52. Dekant W, Kehrer JP. Scientifically unfounded precaution drives European Commission's recommendations on EDC regulation, while defying common sense, well-established science and risk assessment principles. Toxicology Letters. 2013;223(2):A1-A4.
53. Yang RHS. Some critical issues and concerns related to research advances on the toxicology of chemical mixtures. Environmental Health Perspectives. 1998;106(Suppl. 4):1059-63.
54. Altenburger R, Scholz S, Schmitt-Jansen M, Busch W, Escher BI. Mixture toxicity revisited from a toxicogenomic perspective. Environmental Science & Technology. 2012;46(5):2508-22.
55. Jonker MJ, Svendsen C, Bedaux JJM, Bongers M, Kammenga JE. Significance testing of synergistic/antagonistic, dose level-dependent, or dose ratio-dependent effects in mixture dose-response analysis. Environmental Toxicology and Chemistry. 2005;24(10):2701-13.

56. Belden JB, Gilliom RJ, Lydy MJ. How well can we predict the toxicity of pesticide mixtures to aquatic life? Integrated Environmental Assessment and Management. 2007;3(3):364-72.
57. Kutz FW, Barnes DG, Bottimore DP, Greim H, Bretthauer EW. The international toxicity equivalency factor (I-TEF) method of risk assessment for complex mixtures of dioxins and related compounds. Chemosphere. 1990;20:751-7.
58. Logan DT, Wilson HT. An ecological risk assessment method for species exposed to contaminant mixtures. Environmental Toxicology and Chemistry. 1995;14(2):351-9.
59. Matsumura F. Toxicology of Pesticides. New York, USA: Plenum Press; 1985. 598 p.
60. Iwasa T, Motoyama N, Ambrose JT, Roe RM. Mechanism for the differential toxicity of neonicotinoid insecticides in the honey bee, Apis mellifera. Crop Protection. 2004;23(5):371-8.
61. Clements WH, Rohr JR. Community responses to contaminants: Using basic ecological principles to predict ecotoxicological effects. Environmental Toxicology and Chemistry. 2009;28(10):1789-800.
62. Perfecto I. Indirect and direct effects in a tropical agroecosystem: The maize-pest-ant system in Nicaragua. Ecology. 1990;71(6):2125-34.
63. Relyea R. A cocktail of contaminants: How mixtures of pesticides at low concentrations affect aquatic communities. Oecologia. 2009;159(2):363-76.
64. Rohr JR, Crumrine PW. Effects of an herbicide and an insecticide on pond community structure and processes. Ecological Applications. 2005;15(4):1135-47.
65. Castro A, Lacerda M, Zanuncio T, Ramalho FdS, Polanczyk R, Serrão J, et al. Effect of the insect growth regulator diflubenzuron on the predator Podisus nigrispinus (Heteroptera: Pentatomidae). Ecotoxicology. 2012;21(1):96-103.
66. Sánchez-Bayo F, Goka K. Ecological effects of the insecticide imidacloprid and a pollutant from antidandruff shampoo in experimental rice fields. Environmental Toxicology and Chemistry. 2006;25(6):1677-87.
67. Thurber DK, McClain WR, Whitmore RC. Indirect effects of gypsy moth defoliation on nest predation. The Journal of Wildlife Management. 1994;58(3):493-500.
68. Beare MH, Parmelee RW, Hendrix PF, Cheng W, Coleman DC, D. A. Crossley J. Microbial and faunal interactions and effects on litter nitrogen and decomposition in agroecosystems. Ecological Monographs. 1992;62(1):569-91.
69. Hallmann CA, Foppen RPB, Van Turnhout CAM, De Kroon H, Jongejans E. Declines in insectivorous birds are associated with high neonicotinoid concentrations. Nature. 2014;511:341-3.
70. Tennekes HA. The Systemic Insecticides: A Disaster in the Making. Zutphen, The Netherlands: ETS Nederland BV; 2010. 72 p.
71. Poulin B. Indirect effects of bioinsecticides on the nontarget fauna: The Camargue experiment calls for future research. Acta Oecologica. 2012;44:28-32.

72. Norris RF, Kogan M. Interactions between weeds, arthropod pests, and their natural enemies in managed ecosystems. Weed Science. 2000;48(1):94-158.

73. Daryanto I. Pesticide management policy in Indonesia. In: Kennedy IR, Skerritt JH, Johnson GI, Highley E, editors. Seeking Agricultural Produce Free of Pesticide Residues. ACIAR Proceedings. 85. Canberra: Australian Centre for International Agricultural Research (ACIAR); 1998. p. 31-6.

74. Cheke RA, Adranyi E, Cox JR, Farman DI, Magoma RN, Mbereki C, et al. Soil contamination and persistence of pollutants following organophosphate sprays and explosions to control red-billed quelea (Quelea quelea). Pest Management Science. 2013;69(3):386-96.

75. Gustafson DI. Groundwater ubiquity score: A simple method for assessing pesticide leachability. Environmental Toxicology and Chemistry. 1989;8(4):339-57.

76. Hernández M, González LM, Oria J, Sánchez R, Arroyo B. Influence of contamination by organochlorine pesticides and polychlorinated biphenyls on the breeding of the Spanish imperial eagle (Aquila adalberti). Environmental Toxicology and Chemistry. 2008;27(2):433-41.

77. Hewitt AJ, Johnson DR, Fish JD, Hermansky CG, Valcore DL. Development of the spray drift task force database for aerial applications. Environmental Toxicology and Chemistry. 2002;21(3):648-58.

78. Woods N, Craig IP, Dorr G, Young B. Spray drift of pesticides arising from aerial application in cotton. Journal of Environmental Quality. 2001;30(3):697-701.

79. Connolly RD, Kennedy IR, Silburn DM, Simpson BW, Freebairn DM. Simulating endosulfan transport in runoff from cotton fields in Australia with the GLEAMS model. Journal of Environmental Quality. 2001;30:702-13.

80. Luo Y, Spurlock F, Deng X, Gill S, Goh K. Use-exposure relationships of pesticides for aquatic risk assessment. PloS One. 2011;6(4):e18234.

81. Watanabe H, Takagi K. A simulation model for predicting pesticide concentrations in paddy water and surface soil. II. Model validation and application. Environmental Technology. 2000;21:1393-404.

82. Mackay D. Multimedia Environmental Models: The Fugacity Approach. 2nd ed. Boca Raton, FL: Lewis Publishers; 2001. 261 p.

83. Shi Y, Burns M, Ritchie RJ, Crossan A, Kennedy IR. Probabilistic risk assessment of diuron and prometryn in the Gwydir River catchment, Australia, with the input of a novel bioassay based on algal growth. Ecotoxicology and Environmental Safety. 2014;106:213-9.

84. Rose RM, Warne MSJ, Lim RP. Food concentration affects the life history response of Ceriodaphnia cf. dubia to chemicals with different mechanisms of action. Ecotoxicology and Environmental Safety. 2002;51:106-14.

85. Driver C, Ligotke M, Van Voris P, McVeety B, Greenspan B, Drown D. Routes of uptake and their relative contribution to the toxicologic response of northern

bobwhite (Colinus virginianus) to an organophosphate pesticide. Environmental Toxicology and Chemistry. 1991;10(1):21-33.

86. Walker CH. The use of biomarkers to measure the interactive effects of chemicals. Ecotoxicology and Environmental Safety. 1998;40(1-2):65-70.
87. Greig-Smith PW. Use of cholinesterase measurements in surveillance of wildlife poisoning in farmland. In: Mineau P, editor. Cholinesterase Inhibiting Insecticides, their Impact on Wildlife and the Environment. Amsterdam: Elsevier; 1991. p. 127-50.
88. Walker CH. Biochemical biomarkers in ecotoxicology - Some recent developments. Science of the Total Environment. 1995;171(1-3):189-95.
89. Brooks S, Lyons B, Goodsir F, Bignell J, Thain J. Biomarker responses in mussels, an integrated approach to biological effects measurements. Journal of Toxicology and Environmental Health Part A. 2009;72(3-4):196-208.
90. Sánchez-Bayo F, Ahmad R, Goka K. Evaluation of the standard quotient and EcoRR methodologies based on field monitoring from rice fields. In: Kennedy IR, Solomon KR, Gee SJ, Crossan AN, Wang S, Sánchez-Bayo F, editors. Rational Environmental Management of Agrochemicals. ACS Symposium Series. 966. Washington, DC: American Chemical Society; 2007. p. 66-86.
91. Roessink I, Merga LB, Zweers HJ, Van den Brink PJ. The neonicotinoid imidacloprid shows high chronic toxicity to mayfly nymphs. Environmental Toxicology and Chemistry. 2013;32(5):1096-100.
92. Posthuma L, II GWS, Traas TP, editors. Species Sensitivity Distributions in Ecotoxicology. Boca Raton, FL: CRC Press LLC; 2002.
93. Solomon KR, Baker DB, Richards RP, Dixon KR, Klaine SJ, La Point TW, et al. Ecological risk assessment of atrazine in North American surface waters. Environmental Toxicology and Chemistry. 1996;15(1):31-76.
94. Poletika N, Woodburn K, Henry K. An ecological risk assessment for chlorpyrifos in an agriculturally dominated tributary of the San Joaquin River. Risk Analysis. 2002;22(2):291-308.
95. Solomon KR, Giddings JM, Maund SJ. Probabilistic risk assessment of cotton pyrethroids: I. Distributional analyses of laboratory aquatic toxicity data. Environmental Toxicology and Chemistry. 2001;20(3):652-9.
96. Rand G, Carriger J, Gardinali P, Castro J. Endosulfan and its metabolite, endosulfan sulfate, in freshwater ecosystems of South Florida: A probabilistic aquatic ecological risk assessment. Ecotoxicology. 2010;19(5):879.
97. George TK, Liber K, Solomon KR, Sibley PK. Assessment of the probabilistic ecological risk assessment-toxic equivalent combination approach for evaluating pesticide mixture toxicity to zooplankton in outdoor microcosms Archives of Environmental Contamination and Toxicology. 2003;45(4):453-61.
98. Anzecc, Armcanz. Australian and New Zealand Guidelines for Fresh and Marine Water Quality. National Water Quality Management Strategy - Paper No 4.

Canberra: Australian and New Zealand Environment and Conservation Council & Agriculture and Resource Management Council of Australia and New Zealand; 2000.

99. Wang Y, Yu R, Zhao X, An X, Chen L, Wu C, et al. Acute toxicity and safety evaluation of neonicotinoids and macrocyclic lactiones to adult wasps of four Trichogramma species (Hymenoptera: Trichogrammidae). Acta Entomologica Sinica. 2012;55(1):36-45.

100. Stehle S, Knabel A, Schulz R. Probabilistic risk assessment of insecticide concentrations in agricultural surface waters: A critical appraisal. Environmental Monitoring and Assessment. 2013;185(8):6295-310.

101. Giesy JP, Dobson S, Solomon KR. Ecotoxicological risk assessment for Roundup herbicide. Reviews of Environmental Contamination and Toxicology. 2000;167:35-120.

102. Schäfer RB, Paschke A, Liess M. Aquatic passive sampling of a short-term thiacloprid pulse with the Chemcatcher: Impact of biofouling and use of a diffusion-limiting membrane on the sampling rate. Journal of Chromatography A. 2008;1203(1):1-6.

103. Pozo K, Harner T, Shoeib M, Urrutia R, Barra R, Parra O, et al. Passive-sampler derived air concentrations of persistent organic pollutants on a north-south transect in Chile. Environmental Science & Technology. 2004;38(24):6529-37.

104. Metcalf RL. Insecticides in pest management. In: Metcalf RL, Luckmann W, editors. Introduction to Insect Pest Management: Wiley; 1975. p. 235-73.

105. Kookana RS, Correll RL, Miller RB. Pesticide Impact Rating Index - A pesticide risk indicator for water quality. Water, Air & Soil Pollution. 2005;5:45-65.

106. Sánchez-Bayo F, Baskaran S, Kennedy IR. Ecological Relative Risk (EcoRR): Another approach for risk assessment of pesticides in agriculture. Agriculture, Ecosystems & Environment. 2002;91:37-57.

107. Mineau P, Harding KM, Whiteside M, Fletcher MR, Garthwaite D, Knopper LD. Using reports of bee mortality in the field to calibrate laboratory-derived pesticide risk indices. Environmental Entomology. 2008;37(2):546-54.

108. Levitan L, Merwin I, Kovach J. Assessing the relative environmental impacts of agricultural pesticides: The quest for a holistic method. Agriculture, Ecosystems & Environment. 1995;55:153-68.

109. Maud J, Edwards-Jones G, Quin F. Comparative evaluation of pesticide risk indices for policy development and assessment in the United Kingdom. Agriculture, Ecosystems & Environment. 2001;86(1):59-73.

110. Kenaga EE. Predictability of chronic toxicity from acute toxicity of chemicals in fish and aquatic invertebrates. Environmental Toxicology and Chemistry. 1982;1(4):347-58.

111. Barron MG, Raimondo S, Russom C, Vivian DN, Yee SH. Accuracy of chronic aquatic toxicity estimates determined from acute toxicity data and two time-response models. Environmental Toxicology and Chemistry. 2008;27(10):2196-205.

112.Sánchez-Bayo F. From simple toxicological models to prediction of toxic effects in time. Ecotoxicology. 2009;18(3):343-54.

113.Tennekes HA. The significance of the Druckrey-Küpfmüller equation for risk assessment - The toxicity of neonicotinoid insecticides to arthropods is reinforced by exposure time. Toxicology. 2010;276(1):1-4.

114.Druckrey H. Quantitative Grundlagen der Krebserzeugung. Klinische Wochenschriften. 1943;22:532.

115.Sánchez-Bayo F, Goka K. Pesticide residues and bees – A risk assessment. PLoS One. 2014;9(4):e94482.

116.Fenner K, Kooijman C, Scheringer M, Hungerbhler K. Including transformation products into the risk assessment for chemicals: The case of nonylphenol ethoxylate usage in Switzerland. Environmental Science & Technology. 2002;36(6):1147-54.

117.Schenker U, Scheringer M, Hungerbuhler K. Including degradation products of persistent organic pollutants in a global multi-media box model. Environmental Science and Pollution Research International. 2007;14(3):145-52.

118.Fenoll J, Ruiz E, Flores P, Hellín P, Navarro S. Reduction of the movement and persistence of pesticides in soil through common agronomic practices. Chemosphere. 2011;85(8):1375-82.

119.Siebers J, Binner R, Wittich K-P. Investigation on downwind short-range transport of pesticides after application in agricultural crops. Chemosphere. 2003;51(5):397-407.

120.Kennedy IR, Sánchez-Bayo F, Kimber SW, Hugo L, Ahmad N. Off-site movement of endosulfan from irrigated cotton in New South Wales. Journal of Environmental Quality. 2001;30(3):683-96.

121.Wauchope RD. The pesticide content of surface water draining from agricultural fields: A review. Journal of Environmental Quality. 1978;7(4):459-72.

122.Schäfer RB, Paschke A, Vrana B, Mueller R, Liess M. Performance of the Chemcatcher® passive sampler when used to monitor 10 polar and semi-polar pesticides in 16 Central European streams, and comparison with two other sampling methods. Water Research. 2008;42(10-11):2707-17.

123.Mineau P, Whiteside M. Lethal risk to birds from insecticide use in the United States - A spatial and temporal analysis. Environmental Toxicology and Chemistry. 2006;25(5):1214-22.

124.Wehling M, Ohe Wvd, Brasse D, Forster R. Colony losses - Interactions of plant protection products and other factors. Julius-Kuhn-Archiv. 2009(423):153-4.

125.Schmuck R, Schöning R, Stork A, Schramel O. Risk posed to honeybees (Apis mellifera L, Hymenoptera) by an imidacloprid seed dressing of sunflowers. Pest Management Science. 2001;57(3):225-38.

126.Pilling E, Campbell P, Coulson M, Ruddle N, Tornier I. A four-year field program investigating long-term effects of repeated exposure of honey bee colonies to flowering crops treated with thiamethoxam. PLoS One. 2014;8(10):e77193.

127.Henry Ml, Beguin M, Requier F, Rollin O, Odoux JFo, Aupinel P, et al. A common pesticide decreases foraging success and survival in honey bees. Science. 2012;336:348-50.

128.Decourtye A, Devillers J, Aupinel P, Brun F, Bagnis C, Fourrier J, et al. Honeybee tracking with microchips: A new methodology to measure the effects of pesticides. Ecotoxicology. 2011;20(2):429-37.

129.Feltham H, Park K, Goulson D. Field realistic doses of pesticide imidacloprid reduce bumblebee pollen foraging efficiency. Ecotoxicology. 2014;23(3):317-23.

130.Schneider CW, Tautz J, Grünewald B, Fuchs S. RFID tracking of sublethal effects of two neonicotinoid insecticides on the foraging behavior of Apis mellifera. PLoS One. 2012;7(1):e30023.

131.Tremolada P, Mazzoleni M, Saliu F, Colombo M, Vighi M. Field trial for evaluating the effects on honeybees of corn sown using cruiser® and Celest® treated seeds. Bulletin of Environmental Contamination and Toxicology. 2010;85(3):229-34.

132.Goulson D. Neonicotinoids impact bumblebee colony fitness in the field; a reanalysis of the UK's Food & Environment Research Agency 2012 experiment. PeerJ. 2015;3:e854.

133.Von der Ohe PC, Pruss A, Schäfer RB, Liess M, De Deckere E, Brack W. Water quality indices across Europe - A comparison of the good ecological status of five river basins. Journal of Environmental Monitoring. 2007;9(9):970-8.

134.Traas TP, Janse JH, Van den Brink PJ, Brock TCM, Aldenberg T. A freshwater food web model for the combined effects of nutrients and insecticide stress and subsequent recovery. Environmental Toxicology and Chemistry. 2004;23(2):521-9.

135.Schäfer RB, Van den Brink PJ, Liess M. Impacts of pesticides on freshwater ecosystems. In: Sánchez-Bayo F, Van den Brink PJ, Mann R, editors. Ecological Impacts of Toxic Chemicals. Online: Bentham Science Publishers; 2011. p. 111-37.

136.Hose GC, Van den Brink PJ. Confirming the species-sensitivity distribution concept for endosulfan using laboratory, mesocosm, and field data. Archives of Environmental Contamination and Toxicology. 2004;47(4):511-20.

137.Sánchez-Bayo F, Goka K. Evaluation of suitable endpoints for assessing the impacts of toxicants at the community level. Ecotoxicology. 2012;21(3):667–80.

138.Van den Brink PJ, Ter Braak CJF. Principal response curves: Analysis of time-dependent multivariate responses of biological community to stress. Environmental Toxicology and Chemistry. 1999;18(2):138-48.

139.Brock TCM, Roessink I, Belgers JDM, Bransen F, Maund SJ. Impact of a benzoyl urea insecticide on aquatic macroinvertebrates in ditch mesocosms with and without non-sprayed sections. Environmental Toxicology and Chemistry. 2009;28(10):2191-205.

140.Hayasaka D, Korenaga T, Sánchez-Bayo F, Goka K. Differences in ecological impacts of systemic insecticides with different physicochemical properties on biocenosis of experimental paddy fields. Ecotoxicology. 2012;21(1):191-201.

141.Rohr JR, Kerby JL, Sih A. Community ecology as a framework for predicting contaminant effects. Trends in Ecology & Evolution. 2006;21(11):606-13.

142.Sánchez-Bayo F, Yamashita H, Osaka R, Yoneda M, Goka K. Ecological effects of imidacloprid on arthropod communities in and around a vegetable crop. Journal of Environmental Science and Health Part B. 2007;42(3):279-86.

143.Sullivan TP, Wagner RG, Pitt DG, Lautenschlager RA, Chen DG. Changes in diversity of plant and small mammal communities after herbicide application in sub-boreal spruce forest. Canadian Journal of Forest Research. 1998;28(2):168-77.

144.Kefford BJ, Schäfer RB, Liess M, Goonan P, Metzeling L, Nugegoda D. A similarity-index–based method to estimate chemical concentration limits protective for ecological communities. Environmental Toxicology and Chemistry. 2010;29(9):2123-31.

145.Beketov MA, Kefford BJ, Schäfer RB, Liess M. Pesticides reduce regional biodiversity of stream invertebrates. Proceedings of the National Academy of Sciences. 2013;110(27):11039-43.

146.Boyd CA, Weiler MH, Porter WP. Behavioral and neurochemical changes associated with chronic exposure to low-level concentration of pesticide mixtures. Journal of Toxicology and Environmental Health. 1990;30(3):209-21.

147.Kumar A, Doan H, Barnes M, Chapman JC, Kookana RS. Response and recovery of acetylcholinesterase activity in freshwater shrimp, Paratya australiensis (Decapoda: Atyidae) exposed to selected anti-cholinesterase insecticides. Ecotoxicology and Environmental Safety. 2010;73(7):1503-10.

148.Mineau P, Boag PT, Beninger RJ. Effects of fenitrothion on memory for cache-site locations in black-capped chickadees. Environmental Toxicology and Chemistry. 1994;13(2):281-90.

149.Nauen R, Koob B, Elbert A. Antifeedant effects of sublethal dosages of imidacloprid on Bemisia tabaci. Entomologia Experimentalis et Applicata. 1998;88(3):287-93.

150.Desneux N, Decourtye A, Delpuech J-M. The sublethal effects of pesticides on beneficial arthropods. Annual Review of Entomology. 2007;52:81-106.

151.Peakall DB. DDE-induced eggshell thinning: An environmental detective story. Environmental Reviews. 1993;1:13-20.

152.Sánchez-Bayo F. The trouble with neonicotinoids. Science. 2014;346(6211):806-7.

153.Vandenberg LN, Colborn T, Hayes TB, Heindel JJ, Jacobs DR, Lee D-H, et al. Hormones and endocrine-disrupting chemicals: Low-dose effects and nonmonotonic dose responses. Endocrine Reviews. 2012.

154.Zoeller RT, Brown TR, Doan LL, Gore AC, Skakkebaek NE, Soto AM, et al. Endocrine-disrupting chemicals and public health protection: A statement of principles from The Endocrine Society. Endocrinology. 2012;153(9):4097-110.

155.Storrs SI, Kiesecker JM. Survivorship patterns of larval amphibians exposed to low concentrations of atrazine. Environmental Health Perspectives. 2004;112(10):1054-7.

156.Rhomberg LR, Goodman JE, Haber LT, Dourson M, Andersen ME, Klaunig JE, et al. Linear low-dose extrapolation for noncancer health effects is the exception, not the rule. Critical Reviews in Toxicology. 2011;41(1):1-19.

157.Preston BL, Snell TW. Full life-cycle toxicity assessment using rotifer resting egg production: Implications for ecological risk assessment. Environmental Pollution. 2001;114(3):399-406.

158.Tanaka Y, Nakanishi J. Ecological risk estimation based on life table evaluation of chronic toxicity. Mizu Kankyou Gakkaishi. 1998;21(9):589-95.

159.Barnthouse LW, Suter GW, Rosen AE, Beauchamp JJ. Estimating responses of fish populations to toxic contaminants. Environmental Toxicology and Chemistry. 1987;6(10):811-24.

160.Forbes VE, Calow P. Population growth rate as a basis for ecological risk assessment of toxic chemicals. Philosophical Transactions of the Royal Society of London B Biological Sciences. 2002;357(1425):1299-306.

161.Walthall WK, Stark JD. A comparison of acute mortality and population growth rate as endpoints of toxicological effect. Ecotoxicology and Environmental Safety. 1997;37(1):45-52.

162.Tanaka Y, Nakanishi J. Mean extinction time of populations under toxicant stress and ecological risk assessment. Environmental Toxicology and Chemistry. 2000;19(11):2856-62.

163.Stark JD, Banks JE, Acheampong S. Estimating susceptibility of biological control agents to pesticides: Influence of life history strategies and population structure. Biological Control. 2004;29(3):392-8.

164.Sibly RM, Akçakaya HR, Topping CJ, O'Connor RJ. Population-level assessment of risks of pesticides to birds and mammals in the UK. Ecotoxicology. 2005;14(8):863-76.

165.Sibly RM, Newton I, Walker CH. Effects of dieldrin on population growth rates of sparrowhawks 1963–1986. Journal of Applied Ecology. 2000;37(3):540-6.

166.Khoury DS, Myerscough MR, Barron AB. A quantitative model of honey bee colony population dynamics. PLoS One. 2012;6(4):e18491.

167.Roelofs W, Croker DR, Shore RF, Moore DRJ, Smith GC, Akcakaya HR, et al. Case study Part 2: Probabilistic modelling of long-term effects of pesticides on individual breeding success in birds and mammals. Ecotoxicology. 2005;14(8):895-923.

168.Van den Brink PJ, Roelsma J, Van Nes EH, Scheffer M, Brock TCM. PERPEST model, a case-based reasoning approach to predict ecological risks of pesticides. Environmental Toxicology and Chemistry. 2002;21(11):2500-6.

169.Suter GW, Cormier SM. Why care about aquatic insects: Uses, benefits, and services. Integrated Environmental Assessment and Management. 2015;11(2):188-94.

170.Devine GJ, Furlong MJ. Insecticide use: Contexts and ecological consequences. Agriculture and Human Values. 2007;24:281-306.

171.Sánchez-Bayo F, Hyne RV. Comparison of environmental risks of pesticides between tropical and nontropical regions. Integrated Environmental Assessment and Management. 2011;7(4):577-86.

172.Daam MA, Satapornvanit K, Van den Brink PJ, Nogueira AJA. Sensitivity of macroinvertebrates to carbendazim under semi-field conditions in Thailand: Implications for the use of temperate toxicity data in a tropical risk assessment of fungicides. Chemosphere. 2009;74(9):1187-94.

173.Kwok KW, Leung KM, Lui GS, Chu VK, Lam PK, Morritt D, et al. Comparison of tropical and temperate freshwater animal species' acute sensitivities to chemicals: implications for deriving safe extrapolation factors. Integrated Environmental Assessment and Management. 2007;3:49-67.

174.Moore MT, Kroger R, Jackson CR. The role of aquatic ecosystems in the elimination of pollutants. In: Sánchez-Bayo F, Van den Brink PJ, Mann R, editors. Ecological Impacts of Toxic Chemicals. Online: Bentham Science Publishers; 2011. p. 225-37.

175.Schofield N, Williams A, Holloway R, Pyke B. Minimizing riverine impacts of endosulfan used in cotton farming: A science into practice environmental success story. In: Kennedy IR, Solomon KR, Gee SJ, Crossan AN, Wang S, Sánchez-Bayo F, editors. Rational Environmental Management of Agrochemicals. ACS Symposium Series. 966. Washington, DC: American Chemical Society; 2007. p. 360-75.

176.De Snoo GR. Unsprayed field margins: Effects on environment, biodiversity and agricultural practice. Landscape and Urban Planning. 1999;46:151-60.

177.Snelder DJ, Masipiqueña MD, De Snoo GR. Risk assessment of pesticide usage by smallholder farmers in the Cagayan Valley (Philippines). Crop Protection. 2008;27(3-5):747-62.

2

Environmental Pesticides and Heavy Metals - Role in Breast Cancer

1. Introduction

Agriculture chemicals, otherwise referred to as "agrochemicals, " are a large family of chemicals that cover many pest issues associated with farming. For nearly 5, 000 years, crops have been protected by some form of "pesticide." Some of the earliest recorded use of "pesticides" was nearly 5, 000 years ago and involved the use of sulfur dusting in the area of modern-day Iraq and surrounding lands. Ancient Sanskrit hymns (Rigveda) allude to the use of various plant-derived compounds, some of which are poisonous, that can be applied to crops, killing insects yet leaving the crops intact – the first insecticides [1]. There was little advancement in the field of pesticides for thousands of years. In the 1400s, the use of chemicals was tried by farmers to kill various crop-related insects. In most cases, the active ingredient of these chemicals was rooted in the actions of heavy metals (arsenic, mercury, and lead). Two hundred years later, the alkaloid, nicotine, was being investigated as a potential agent to eliminate insects from crops. The utility of nicotine prompted additional investigation into the use of natural products as insecticide/pesticide agents. Interestingly, two of the compounds which were developed, pyrethrum and rotenone, are still used today experimentally for various research-related purposes. The use of arsenic-based compounds prevailed for hundreds of years until the mid-1900s as not only agents for pest prevention, but also for poisonings [2]. A shift in the pesticide industry began to occur in the early- to mid-1900s. This began with the identification of dichlorodiphenyltrichloroethane (DDT) as a potent insecticide. DDT was originally developed 75 years earlier, but it was not until nearly 1940 when Dr. Paul Müller discovered that DDT was a very effective insecticide (later being awarded the Nobel Prize in Physiology or Medicine in 1948 for this discovery). Pesticides that contained chlorine were initially referred to as "organochlorines, " with DDT being considered the prototype organochlorine, which many subsequent agents were based on. Very quickly, considering the timeline of pesticide development and use over thousands of years, organochlorine pesticides were banned in the United States and replaced by newer pesticide derivatives. By substituting a phosphate for chlorine, the "organophosphate" class of pesticides was developed. In addition, the carbamates class was also developed and the organophosphates and carbamates effectively replaced organochlorine pesticides by 1975. Since then, pyrethrin compounds have become the dominant insecticide. Herbicides became common in the 1960s, led by "triazine and other nitrogen-based compounds, carboxylic acids such as 2, 4-dichlorophenoxyacetic acid, and glyphosate" [2].

In general, these pesticides can broadly include compounds/chemicals that can kill insects, invasive plants, invasive other organisms – generally anything that will kill pests that damage farmland. The drawback to the use of these chemicals is that they are all toxic and in most cases can enter the groundwater, or the food chain. Even if handled and stored properly, the chance of accidental exposure is greatest for those who live closer to the areas being treated. Normally, the release of toxic/ dangerous chemicals into the environment is tightly regulated and controlled. The one unique aspect of pesticide use is that these are highly toxic chemicals and are voluntarily applied in the environment. By definition, a pesticide is a compound/ chemical that will kill a variety of pests and the more specific name will indicate what is being killed, such as herbicides (weeds), fungicides (fungi), insecticides (insects), rodenticides (rodents), etc. In many instances, we can generalize and simply refer to any of these agents as a pesticide. Because of their widespread use, ease of accessibility, and, in some instances, low cost, the impact of pesticides on the environment has become profound and significant. This impact is no longer confined to agricultural fields, but can also include, homes, businesses, academic institutions, and some areas otherwise believed to be protected. The link between pesticides and cancer has long been a concern. Besides low-level accidental exposure, large-scale contamination can occur when there are industrial accidents or waste dumping of large quantities of raw agrochemicals. The production of agrochemicals is still big business, with the top producers reporting revenue of nearly $15 billion in 2013. Yet, in 2005 it was reported that the economic impact on pesticide use in the United States was nearly $10 billion [3]. These studies broke down the pesticide cost on U.S. economy as follows: $1.1 billion for public health, $1.5 billion for pesticide resistance, $1.4 billion crop losses due to pesticides, $2.2 billion in avian losses, $2 billion in groundwater contamination. The ratio of dollars spent on pesticides to dollars for crops saved is about 1:4 [4]. This ratio is essentially a return on an investment. You can expect nearly a fourfold return on your expenditures due to increased crop yield (due to reduced pest burden). A return this large is a significant benefit to the growers who can grow a more diverse annual crop, and will generate more crop output on an annual basis. This increased ability to generate goods would translate into a benefit for the consumer [5]. Alternatives to pesticide use have been explored and are currently being developed. Some of these alternative methods involve changes in cultivation methods, use of biologically derived pesticides (extracts or other compounds of organism that may kill the pest in question). There has been some indication that alternative methods can be equally efficacious to chemical pesticides and are safer to human populations also. The loss of crops and the economic impact on widespread damage resulting from pest infestation would be significant and catastrophic. Therefore, although agrochemical use is vital and necessary, it does come with a relatively high cost.

Although there have been significant inroads in the fight against breast cancer since 1994, there is still much to be done and even more that needs to be understood. Still, nearly 12% of women will experience invasive breast cancer in their lifetime – that is, over 20 million cases. It is understood that there are multiple factors that underlie the pathogenesis of breast cancer, either invasive or noninvasive. Key factors include 1) genetic predisposition, 2) environmental factors, and 3) lifestyle. None of the factors are exclusive, but rather inclusive of each other. Studies have demonstrated that there are numerous synthetic chemicals that can mimic the actions of estrogen. Exposure to estrogen like compounds can come from many sources, including cosmetics and other household items [6]. Not all are detrimental, such as the phytoestrogens which are derived from plant sources. But there are compounds referred to as xenoestrogens, or "endocrine disruptors." Regardless, they include numerous pesticides, including the broad class of organochlorine pesticides. These compounds have demonstrated adverse medical outcomes [7]. In addition, there are certain metals which are found in the environment that have displayed estrogenic activity [8]. Cadmium-based chemicals have been shown to be highly toxic and carcinogenic [9]. Ultimately, the goal for treatment and improved prognosis is the development of compounds which can counteract, or reverse, the effects seen after exposure to these xenoestrogen environmental contaminants. Many of the available compounds focus on estrogen-dependent tumors and the response to these drugs is favorable. But tumors lacking the estrogen receptor respond poorly to current treatments. The literature is lacking detailed studies on the coexposure of OCPs and metals and how these complex exposures may alter tumor development. Recent studies and reviews have only begun to shed light on this situation [10, 11].

The prefix "xeno" refers to something with a different origin. Therefore, the term xenoestrogen is a form of our endogenous estrogen hormone that imitates estrogen but comes from a nonbody source. Xenoestrogens are also called "environmental hormones" or "endocrine disrupting compounds" (EDC). These forms of estrogen may be clinically significant because they can mimic the effects of endogenous estrogen and have been theorized to be involved in numerous pathologies. Xenoestrogens can be either synthetic or natural chemical compounds. Synthetic xenoestrogens are widely used industrial compounds (Table 1), such as PCBs, BPA, and phthalates, which have estrogenic effects on a living organism even though they differ chemically from the normal estrogenic hormones (estradiol, estriol, estrone). Naturally occurring xenoestrogens include plant-derived phytoestrogens. Because the primary route of exposure to these compounds is by consumption of phytoestrogenic plants, they are sometimes called "dietary estrogens." Mycoestrogens, estrogenic substances from fungi, are another type of xenoestrogen that are also considered mycotoxins. An increasing amount of xenoestrogenic compounds have been deposited into the environment over the last century. The potential ecological and human health impact of environmental xenoestrogens has been of increasing concern over the last 20-30 years.

Xenoestrogens have been implicated in a variety of medical pathologies, and over the last decade numerous studies have found evidence of adverse effects on human and animal health [12-15]. There is a concern that xenoestrogens may act as false messengers and disrupt the process of puberty and reproduction. The induction of cytochrome P450 isozyme, CYP1A, has been established to be a good bioindicator for xenoestrogen exposure. Another potential effect of xenoestrogens is on oncogenes, specifically in relation to breast cancer. The impact of xenoestrogens on breast tumor growth has been equivocal, leading to some speculation that these agents may not be involved in breast cancer progression. It does appear to depend on the form of cancer and the type of xenoestrogen exposure. However, there is growing evidence that xenoestrogens can increase breast cancer growth in tissue culture. One complication in drawing an accurate conclusion is that many studies have simply examined high concentrations/doses in short exposure periods. In most situations, exposures in this fashion would not occur unless exposed in a manufacturing or industrial setting. Also, the majority of the population will not be exposed to singular compounds. There will be "cocktails" of sorts, mixtures of many different xenochemicals with exposures to pesticides as well as heavy metals being likely. Additional work needs to be done before more accurate conclusions can be drawn regarding the effects of either the xenoestrogens or metalloestrogens (a form of xenoestrogen).

Table 1. Grouping of xenoestrogens into four major classes

Estrogenic Class	Subclass	Representative Compounds
Phytoestrogens	Flavanones	Hesperetin Naringenin Liquiritigenin Pinocembrin
	Flavones	Acacetin 7, 8-dihydroxyflavone Mirificin
	Prenylflavonoids	8-Prenylnaringenin 6-Prenylnaringenin 8-Geranylnaringenin 6, 8-Diprenylnaringenin Icaritin
	Isoflavones	Daidzein Daidzin Genistein Genistin Glycitein Ononin Pruentin Formononetin Peurarin Tectorigenin Biochanin A
	Isoflavanes	Equol (S-Equol) Glabridin
	Dihydrochalcones	Phloretin Phlorizin
	Coumestans	Coumestan Coumestrol Desmethylwedelolactone Wedelolactone
	Lignans	Enterodiol Enterolactone
	Flavonolignans	Silybin Cinchonain – Ib

Estrogenic Class	**Subclass**	**Representative Compounds**			
	Flavonols	Quercetin			
	Others	Deoxymiroestrol Miroestrol Resveratrol α-Isosparteine			
Mycoestrogens		Zearalanone Zearalenone α-Zearalanol α-Zearalenol			
Synthetic		4-MBC Alkylphenols Atrazine BHA BPA	DDE DDT Dieldrin Dioxins Endosulfan	Erythrosine Heptachlor Lindane Methoxychlor Parabens	PBB PCB Phthalates Propyl gallate
Metalloestrogens		Aluminum Antimony Arsenite Barium Cadmium	Chromium (II) Cobalt Copper Lead Mercury		Nickel Selenite Tin vanadate

4-MBC - 4-Methylbenzylidene camphor; BHA - Butylated hydroxyanisole; BPA - Bisphenol A; PBB - polybrominated biphenyl; PCB - polychlorinated biphenyl

Below is additional description of synthetic xenoestrogens and their current use [16-18].

- Atrazine: widely used as a herbicide to control broad-leaf weed species and is the second largest seller in the United States. Atrazine is also applied to Christmas trees, lawns, golf courses, and other recreational areas.
- BPA (Bisphenol A): monomer used to manufacture polycarbonate plastic and epoxy resins used as a lining in most food and beverage cans in excess of 6.4 billion lbs/year. BPA can hydrolyze and the leaching of BPA has led to widespread human exposure.
- DDT (Dichlorodiphenyltrichloroethane): banned in 1972 due to its hazardous effects on the environment, it continues to be used in many parts of the world. DDT and its metabolites DDE and DDD are persistent in the environment and accumulate in fatty tissues.
- Dioxin: released during combustion processes, pesticide manufacturing, and chlorine bleaching of wood pulp. Consumption of animal fats is thought to be the primary pathway for human exposure.
- Endosulfan: is an insecticide used on numerous vegetables, fruits, cereal grains, and trees. Human exposure occurs through food consumption or ground and surface water contamination.
- PBB (Polybrominated biphenyls): used in computer monitors, televisions, textiles, and plastics foams to make them more difficult to burn. Production stopped in the mid-1970s, but due to the very slow degradation, PBBs can still be detected in moderate concentrations in the environment.
- PCBs (Polychlorinated biphenyls): also known as chlorinated hydrocarbons. PCBs were used as insulating fluids and coolants. PCBs were banned in 1979 but like DDT continue to persist in the environment.
- Phthalates: plasticizers which increase durability and flexibility to plastics. Both high and low molecular weight phthalates are used in a variety of consumer products which can include perfumes, lotions, cosmetics, varnishes, lacquers, and coatings including timed releases in pharmaceuticals.
- Zeranol: anabolic growth promoter for livestock, banned in Europe, but still present as a contaminant.
- There are over 25 other miscellaneous xenoestrogens that range from banned insecticides that may still be stored onsite [DDT, Dieldrin, endosulfan, methoxychlor], to "normal" household products like 4-methylbenzylidene [suntan lotion], to estrogen compounds like ethinylestradiol [birth control pills] plus a host of others. Collectively, even in small concentrations, these agents may have localized significant effects.

2. Pesticides

What is a pesticide? As defined by the Food and Agriculture Organization of the United Nations, a pesticide is: "*any substance or mixture of substances intended for preventing, destroying or controlling any pest, including vectors of human or animal disease, unwanted species of plants or animals causing harm during or otherwise interfering with the production, processing, storage, transport or marketing of food, agricultural commodities, wood and wood products or animal feedstuffs, or substances which may be administered to animals for the control of insects, arachnids or other pests in or on their bodies*" [FAO, http://www.fao.org/docrep/x1531e/X1531e02.htm]. By definition, classification of a pesticide is an exceptionally broad term and encompasses a multitude of different commercial and/or private uses. Due to the diverse nature of pesticides, they have been linked to a wide range of human health issues. The role of pesticides in the development of cancers, alterations in reproduction, as well as an array of symptomologies associated with acute, or short-term, exposures.

Comparison between short-term (acute) and long-term (chronic) exposure:

- Acute exposure: General irritation to the contact areas such as skin and eyes. If absorbed or inhaled, there may be peripheral nerve damage (neuropathies), or potential effects on the brain, which will result in headaches, dizziness, and fatigue. Exposure to significantly large quantities of a pesticide may result in a systemic poisoning that results in generalized organ failure and, potentially, death.
- Chronic exposure: Chronic exposure is occasionally the most difficult to diagnose and assess. Especially if the individual was exposed to low levels of the pesticide. Initially, there may be no symptomology, or visible signs that a person was exposed. With a slow degeneration of a particular physiological response, symptoms can become visible weeks, months, or even years after the exposure has ended. In some instances, pesticides that have been shown to bioaccumulate, can concentrate within the body and over time will elicit a toxic response. Pesticides can cause many types of cancer in humans. Some of the most prevalent forms include leukemia, non-Hodgkin's lymphoma, brain, bone, and breast, ovarian, prostate, testicular, and liver cancers.

Since the 1980s, increasing amounts of data have indicated there are both direct and indirect roles for pesticides in the disruption of the mammalian endocrine system. Alterations in normal homeostatic endocrine functioning due to pesticide exposure occurring from fetal development through adulthood can affect the delicate balance of hormonal systems involved with normal development. Within the larger universe of pesticides, a subset has been identified that exhibits abilities to alter mammalian (and nonmammalian) endocrine systems. Collectively, these agents are referred to as "endocrine disruptors." There are nonpesticide chemicals that are included on this list, such as the metalloestrogens, but pesticides on the list of endocrine disruptors include DDT, lindane, parathion, plus a multiple of others that are chemically and structurally similar.

The young and elderly are particularly sensitive to the actions of pesticides. In particular, children under the age of 15 exhibit significant risk due to the continuing development of the central nervous system. In addition, the child's developing respiratory, gastrointestinal, hepatic systems can be likely targets for pesticide toxicity. Compounding the potential problem is the child's close proximity to where the pesticides are located. In most cases, pesticide storage will be at ground level. Also, application of pesticides to crops, plants, grasses, etc., puts the pesticide at locations where the child may physically come into contact with the pesticide through normal activities. This potential elevated risk factor, coupled with the child's inability to eliminate or neutralize the pesticide can result in numerous physiological deficits observed during development, suggesting that children are suffering disproportionately from the impact of pesticide exposure.

Pesticides can be classified by target organism (herbs/weeds, insects, fungi, rodents), chemical structure (organic, inorganic, synthetic, or biological/ biopesticides), and physical state (powder, solid, liquid or gaseous). Plant-derived pesticides, or "botanicals, " have been developing rapidly. These include the pyrethroids (plant-derived pesticides), rotenoids, nicotinoids, and a fourth group that includes strychnine/scilliroside. The broad classes of chemical pesticides can include:

- Organophosphates (OPs): affect the nervous system by disrupting the enzyme acetylcholinesterase, which is responsible for the breakdown, or degradation, of acetylcholine. Inhibition of this enzyme will lead to increased parasympathetic tone, due to increased activity at both muscarinic and nicotinic cholinergic receptors. They were developed during the early 19th century, but their effects on insects, which are similar to their effects on humans, were discovered in 1932. Some are very poisonous. Individuals exposed to high levels of OPs can develop what is referred to as "acute cholinergic syndrome, " which is characterized by a variety of symptoms including rhinorrhea, salivation, lachrymation, tachycardia, headache, convulsions, and death [19]. However, they usually are not persistent in the environment so the threat of bioaccumulation is not as great compared to organochlorine pesticides. The more toxic organophosphates have been replaced by the less-toxic carbamates.
- Carbamates: affect the nervous system by disrupting the enzyme acetylcholinesterase, which is responsible for the breakdown, or degradation, of acetylcholine. Inhibition of this enzyme will lead to increased parasympathetic tone, due to increased activity at both muscarinic and nicotinic cholinergic receptors. The enzyme effects are usually reversible. There are several subgroups

within the carbamates family including the thiocarbamate and dithiocarbamate subclasses.

- Organochlorines: A major physiological action of organochlorines is the disruption of the sodium/potassium balance of the nerve fiber. Altering the resting membrane potential may have two effects; one of which is the ability to turn an excitable cell into the "on" position. In the instance of organochlorine compounds, that results in constant action potentials traveling down the nerve pathways. In general, the organochlorine group has a wide range of toxicities with some agents being relatively nontoxic, yet other compounds eliciting a high level of toxicity and cellular damage. Collectively, the ban on organochlorines is near universal, due to the fact that these agents can persist in the environment for significant lengths of time.
- Pyrethroid: They were developed as a synthetic version of the naturally occurring pesticide pyrethrin, which is found in chrysanthemums. They have been modified to increase their stability in the environment. Some synthetic pyrethroids are toxic to the nervous system.
- Sulfonylureas: The sulfonylurea class prototype was originally developed and used to treat Type II Diabetes, but many of the analogs in the sulfonylurea family can be found commercially available to control weeds. Most of the sulfonylurea agents are broad spectrum and work by inhibiting the enzyme acetolactate synthase which is vital for weed growth. The sulfonylureas are generally safer to humans compared to the other classes of agents, and only a fraction of the sulfonylurea agent is needed compared to other herbicides. Due to the highly toxic nature of many of these compounds, most have been discontinued in the United States, yet, there are significant stock piles that are in need of disposal, or are still extensively used outside of the United States. For this chapter, most of the discussion will focus on the actions of the organophosphate and organochlorine agents.

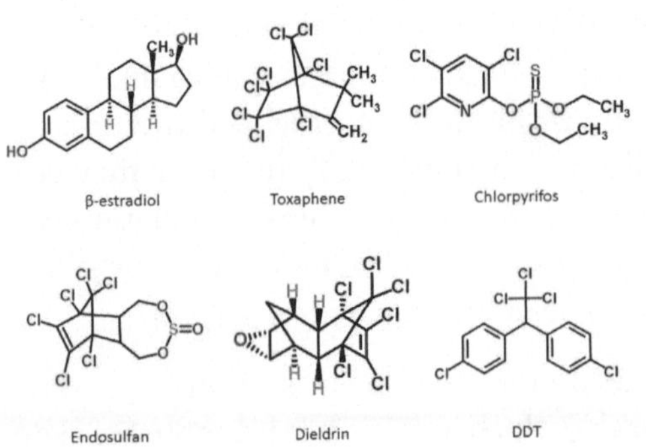

Figure 1. Chemical construction comparison between β-estradiol and pesticides with known or suspected estrogenic activity. Figures all obtained from ChemSpider [www.chemspider.com]

2.1. Environmental impact

In addition to the desired effects of pesticides, such as killing invasive weeds, crop-damaging insects, etc., most pesticides have demonstrated the ability to persist in the environment for extended periods of time. Ground applications can result in a leaching effect into the groundwater, and eventually into the water supply. These agents are accumulated and concentrate in the sediment found in lakes, rivers, and streams. This allows introduction into the food chain when ingested by fish. This cycle can then continue in a forward-moving manner. These characteristics of pesticides have made the assessment of the environmental impact very difficult. Years, even decades, after discontinuing the use of a particular pesticide, it may still be detected in the environment at levels that could pose a direct risk to humans or other mammals. There are also potential indirect risks associated with pesticide bioaccumulation. High levels of toxic pesticides can damage the soil, alter the quality of the soil, and potentially reduce crop yield or quality. By impacting food production, certain pesticides may elicit detrimental effects, both direct and indirect, for significant periods of time.

There are thousands of chemicals that are available, yet only a few hundred have undergone toxicity testing. Legislation has been recently introduced to update the Toxic Substance Control Act and increase the testing of these chemicals. The Environmental Protection Agency has published its list of pesticide active ingredients that are to be screened under the Federal Food, Drug and Cosmetic Act [20]. This final list only contains 67 compounds that are considered either active or inert ingredients in pesticide preparations – a number that falls significantly short of the total number of compounds that are currently available. Accompanying this lack of information is the uncertainty as to whether chemicals can be causative agents in the development of breast cancer [21]. The broad class of organochlorine compounds includes organochlorine pesticides (OCPs). Although most OCPs have been banned from use in the United States, with dichlorodiphenyltrichloroethane (DDT) being the prototype compound in this class, there are a few still in use. Endosulfan has been linked to numerous pathologies, and is in the process of being phased out by 2016. Lindane and Dicofol are still used in the United States. Although the number of OCPs is dwindling, there is still the potential that stored OCPs may be used. In addition, OCPs have been shown to persist for extended periods of time in the environment [22] and in the body [23, 24]. Collectively, these data would suggest that OCPs may be a persistent health risk for decades to come with regard to long-term toxicity. Synthetic organochlorine agents can be used in the production of plastic (polyvinyl chloride, PVC), chloromethanes (solvents – carbon tetrachloride; and anesthetics – chloroform), insulators (polychlorinated biphenyls – PCBs) on electrical wire, but these have been phased out. You may still find PCBs in older homes, thus still making PCBs a potential risk factor. With the increased production of hazardous chemicals like heavy metals for other industrial uses as well as the production of these compounds as by-products of industrial processing has created a series of unique challenges for the disposal of these compounds. A variety of heavy metal combinations that were released into waste effluent has demonstrated a synergistic lethality on aquatic organisms [25]. Based on this synergism, the conventional mechanisms of determining lethality (LD50, etc.) may not be accurate. Reductions in toxin content to 1/3 or 1/6 of LD50 for a single toxin may not be a large enough dilution to adequately reduce toxicity in aquatic environments [25]. There have been numerous reports of contamination worldwide, land-based and aquatic-based in the food chain. In the United States, marine mammals near Alaska (walrus, beluga whale, bowhead whale, and rigged seal) have shown increased levels of a variety of toxins such as DDT, PCBs, mercury, and cadmium [26]. In addition, chlordane and toxaphene have been detected in increasing concentrations. The elevations in these compounds, and their estrogenic nature, may interfere with the reproduction of these species, thereby endangering their future. A combination of pesticides and heavy metals has been shown to induce larval abnormalities in Pacific oysters, reducing their ability to grow and proliferate. A trend, if allowed to continue, will result in the loss of this oyster species permanently. On the east coast of the United States, elevated levels of organochlorine pesticides, PAHs, and heavy metals have been found in oysters collected in the marshland [27]. These agents will then enter the food chain and be consumed by other animals or humans further bioaccumulating. Elevated levels of organochlorine pesticides and heavy metals have been found in a species of eel off the coast of Belgium [28]. A variety of edible species in China have been reported to have elevated levels of organochlorine pesticides and heavy metals [29]. Other reports have substantiated these findings and extended them to other parts of the globe [30, 31]. Organochlorine pesticides and heavy metals have become a significant worldwide hazard contaminating numerous species of marine life, posing a threat to the survival of that species as well as becoming an increasingly significant threat for entering the general human population through their consumption. As a result, increased vigilance needs to be implemented to prevent the loss of these species and the further contamination of the human population.

Increased incidence of contamination is not confined to the aquatic environment. Over the past two decades, there has been an increased awareness of potentially toxic compounds entering the food chain. This could be through the spraying of pesticides to maintain crop integrity, or through groundwater contamination following industrial contamination. The presence of pesticides, hormones, food additives, and other chemicals has increased the public's concern over the status of their food supply. Food safety is a major public concern worldwide and food consumption has been identified as the major pathway for human exposure to certain environmental contaminants. "Chemophobia" (the fear of chemicals in foods) is the major reason for individuals to choose organic food – believing that these types of foods would be free of synthetic pesticides. Generally, organic farms avoid the use of chemically derived pesticides, fertilizers, and other additives, but instead favor more "natural" means of removing pests and improving crop yield. Many believe that "organically" grown crops are "better" health-wise. In some instances, this is the case, but even without some of the usual additives, there is an increased contamination when contaminated groundwater is used for irrigation [32]. Even with the differing farming systems, and the public's perception that organically grown foods are healthier, increased surveillance and screening must be done to monitor the amount of potentially harmful chemicals that are being ingested on a daily basis [33]. In Europe, a continent-wide study involving the World Health Organization found that there were dangerous levels of pesticides and heavy metals in foods which were previously believed to be safe [34]. The major difficultly in analyzing field samples for contamination is that multiple other chemicals and compounds can be identified. The complexity of these chemical–chemical interactions within a biological system cannot be understated. It is the complexity of these interactions which obscure the ability to definitely determine and predict the effect of the sample on a biological system. Mansour and Gad [35] have developed an ecological testing system using *Daphnia magna* Straus, which can be performed rapidly and easily by a field technician. One advantage is that it will test the entire sample, without regard for the number of contaminants. It can also be performed without large-scale and expensive equipment. The drawback is that the individual toxin cannot be distinguished [35].

Over-the-counter "natural products", herbal preparations that have reported therapeutic benefits, including herbal medicine such as Saint John's Wort, Ginseng, Gingko, etc. generate significant revenues and their use is increasing, especially in the United States [36, 37]. Spending on these natural products in the United States has more than doubled in the last 10 years to nearly $15 billion in sales [37]. One group reported in the late 1990s that some natural products such as curcumin and isoflavonoids can act as an estrogen receptor antagonist and block the effects of estrogen overstimulation [38]. They found that phytoestrogen-related compounds significantly attenuated the effects seen following exposure to the pesticide DDT, and the pollutants 4-nonylphenol and 4-octylphenol [38]. Several studies have shown that Chinese herbal medicines and other botanical supplements may be contaminated with heavy metals, and in some cases at toxic levels [39-41]. There are differences between the preparations of natural medicines. There is the "raw" form – which usually will take the form of the root, nut, bean, etc. Most concern has been directed toward proprietary blends. At this time, it is believed that the external toxins are introduced [42]. Interestingly, some known toxins, such as arsenic and other forms of the arsenicals, lead and mercury, are intentionally added for their purported medical value [43-45]. Contrary to public perception, heavy metals have been detected in raw preparations and, in fact, some species of plants are known to concentrate (bioaccumulate) heavy metals [42]. Heavy metal contamination has also been found in some supplements sold in the United States, but at levels that were not deemed to be hazardous [46]. Since most over-the-counter preparations and those sold over the Internet, fall under the category of "supplement, " there are no extensive guidelines for the limits of heavy metals and pesticides that can be contained in the preparations. Collectively, it appears that consumption of certain phytoestrogens may have a therapeutic benefit in ameliorating the effects observed following exposure to some pesticides and/or environmental toxins, but with some possible contamination hazard of their own. Although the levels of pesticides and heavy metals that would be ingested by consuming raw, or formulated, herbal preparations would be low in the vast majority of the preparations, care must be used and stringent monitoring must be utilized to prevent accidental intoxication of individuals who are ingesting over-the-counter herbal preparations.

2.2. Pesticides and the link to cancer

While agriculture has traditionally been tied to pesticide-related illnesses, over half of the commonly used pesticides are linked to cancer. The true burden of environmentally induced cancer is greatly underestimated. Chemicals can trigger cancer in a variety of ways, including disrupting hormones, damaging DNA, inflaming tissues, and turning genes on or off. Many pesticides are known to cause cancer, and virtually everyone in the United States is exposed to them on a daily basis. In animal studies, many pesticides are carcinogenic (e.g., organochlorine, creosote, and sulfallate), while others (notably, the organochlorine agents DDT, chlordane, and lindane) are tumor promoters. Some contaminants, such as arsenical compounds, in commercial pesticide formulations also may pose a carcinogenic risk. In humans, arsenic compounds and certain insecticides used occupationally have been classified as carcinogens with organochlorine insecticides being linked with cancers of the lung and breast [47]. With the development of new technologies, the casual association of pesticides with cancer has been strengthened. Most of the human studies have involved retrospective studies examining the incidence of particular cancers like breast cancer and their exposure to pesticides. Now, there is a great deal of research and focus on the identification of biomarkers that can be used to identify these associations between pesticides and cancer. One technology that may be invaluable for biomarker development will be the use of toxicoproteomic-based data [48]. Regardless of the direction that the technology takes us in the future, there is a significant amount of work that still needs to be done to more fully understand how pesticides can be affecting the development of breast cancer as well as other chronic human disorders [49].

Farmers in many countries, including the United States, have lower overall death rates and cancer rates than the general population. Although death rates may be lower, in-depth analysis of the incidence of specific diseases (cancer) indicate that there may be a higher rate of certain cancers among farmers and agricultural workers. Additional work attempting to correlate the types of disease with the chemical exposure has not been unequivocal and there have been many conflicting reports either supporting or refuting the cancer–pesticide link. Similar to what was discussed earlier, the developing young, from fetus to mid-teen, are some of the highest risk groups for disease resulting from pesticide exposure. Young females who may be exposed to an endocrine disrupting agent (such as DDT) have been reported to have a higher incidence of breast cancer compared to populations which were not exposed. Interestingly, this same risk exists if the parents are exposed prior to conception, suggesting that either the pesticide, or changes in the maternal system, can be passed onto the fetus. Collectively, these rural populations of agricultural works and their families tend to be exposed at a higher rate and at higher concentrations than the general population, and as such may experience a higher incidence of particular cancers (breast) than the general population [6].

Yet, to say that there is a significant correlation between pesticide exposure and breast cancer incidence would be an overstatement. Retrospective studies have yielded conflicting results. Most of the studies recognize shortcomings, such as small population size, difficulty in assessing pesticide exposure, and correlating blood pesticide levels to the progression of breast cancer. Pesticides that had widespread use, and were widely popular and commercially available, DDT, DDE, and dieldrin, are the best examined. Since many of these pesticides have exhibited severe health-related adverse effects, older pesticides have been banned, or their use restricted. Yet, due to bioaccumulation, persistence in the environment, and the need to dispose of older stockpiles, they remain an environmental and health concern. As the conflicting evidence has come forward, it appears that one factor that may be important for the toxicity to DDT/DDE/dieldrin is ethnic background. This suggests that genetic polymorphisms between ethnic groups may predispose individuals to cancer risk [50]. A post hoc meta-analysis by Ingber et al. [51] concluded that there was no definite correlation between DDT/DDE exposure and breast cancer. This was done via a PubMed and Web of Science search of nearly 500 cases. Although there were slight elevations in the levels of DDT and the incidence of breast cancer, none of their correlations were statistically significant. Of course, the analysis may have missed some ethnic groups or other populations where a positive correlation existed, but offers a strong conclusion that these compounds alone may not be enough to promote breast cancer development. In an Indian population from Jaipur, women with higher levels of DDT and its metabolites, dieldrin and heptachlor, were correlated with the incidence of breast cancer [52]. A study examining woman of Caribbean descent aimed to correlate pesticide exposure with the incidence of both prostate and breast cancer. Due to relatively high organochlorine pesticide use on the island of Martinique, and what appeared to be an elevated frequency of prostate and breast cancers, the study intended to draw a correlation between pesticide exposure and cancer. They report that there is a positive correlation in the exposure to pesticides and the incidence of breast cancer [53]. Contrary to these studies, research into the causal relationship between pesticides and breast cancer in the United States has not been as convincing and in some instances mixed

[50]. An east coast study examined a population of women from Long Island, New York, and found that there was some evidence for a positive correlation between exposure and cancer. It was not a strong correlation, and involved other factors, but would warrant further investigation [54]. Yet a west coast study, the "California Teachers Study" cohort showed no correlation between pesticide exposure and the incidence of breast cancer [55]. Clearly, there is a need to correlate all of the different factors involved in the pathogenesis of breast cancer with the exposure to suspected carcinogens.

It has been known for some time that select pesticides interfered with the action of estrogen. This term was loosely referred to as an "Endocrine Disruptor." This category has since been expanded to include many items such as detergents, disinfectants, plastics, and an increasing number of pesticides. There are three main actions of an endocrine disruptor. First, the pesticide can mimic the actions of estrogen (or testosterone), thereby causing an increase in estrogen-related physiological responses. Second, the pesticide can act like an antagonist and block the actions of estrogen at its receptors. This will prevent the normal physiological responses associated with estrogen stimulation of its receptor. Third, the pesticide may have a broader effect and interfere with the synthesis, transport, metabolism, or elimination of estrogen. This may have a variable effect with either an increase or decrease in estrogenic effects being observed. Regardless, the normal homeostasis of the system will be disrupted.

Of the broad classes of pesticide – the organochlorine class has been the most extensively studied and has shown to be the most potent as an endocrine disruptor. In the 1990s, studies were done in cell culture model systems which demonstrated that organochlorine pesticides were more potent at activating the estrogen receptor, whereas organophosphorus pesticides were relatively ineffective at modifying estrogen receptor activity [56]. Although the focus of this chapter is breast cancer, other sites of estrogen activity, such as the uterus, also demonstrated that organochlorine pesticides can stimulate receptors on the uterus leading to the development of uterine leiomyoma [57]. The preponderance of data has focused on the ability of these compounds to act directly on estrogen receptors. Initial screening, if originally negative for estrogen receptor activity, does not necessarily mean that organochlorine pesticides are devoid of cancer-stimulating properties. There are indirect mechanisms by which pesticides can influence the proliferation and function of breast tissue. One possible indirect mechanism can be through the binding to tubulin and arrest of the G2/M cycle [58]. A recent study [59] has substantiated the ability of pesticides to influence cellular proliferation. Ventura et al. [59] demonstrated that chlorpyrifos induces a redox imbalance altering the antioxidant defense system and inhibition of cellular proliferation of ERK1/2 phosphorylation. They concluded that the effect on ERK1/2 phosphorylation was not a direct effect but an indirect effect as a result of the changes in the redox state of the cell. Also, most exposures will involve mixtures of compounds. Recent work has begun to explore these effects of mixtures on proliferation of breast tissue. In addition to direct effects of the constituents on the estrogen receptor, the mixtures may also interfere with cellular proliferation [60], inhibition of androgen activity [60], and the upregulation of protein kinase genes associated with tumor development [61]. Not all effects on breast cancer cells are mediated by the genomic actions of estrogen receptors. Activation of CaMKIV pathways is structure-dependent, with estrogen being the most potent activator [62]. Similar results were observed with the activation of PI3-K, MAPK, and PKC. Select phytoestrogens such as resveratrol displayed minimal activity [62]. This would suggest that there is a structural requirement for non-estrogen compounds to elicit effects on intracellular kinase systems. Collectively, the summation of these effects would have a detrimental on cellular function.

The majority of the data in the literature has examined the effects of pesticides on either the estrogen α or β receptor (ERα or ERβ). Structurally, these pesticides can be highly diverse with little resemblance to β-estradiol, the most potent of the various forms of the estrogen hormone (estrone and estriol being the other two). Yet, even with these structural disparities, the pesticides can still function as an endocrine disruptor. In the mid-1990s it was reported that methoxychlor and DDT bound to estrogen receptors and that it was potentially through this action that binding to estrogen receptors in the fetal brain, neural development was altered and as a result, the territorial behavior in male mice was affected [63]. This substantiated an earlier report using cell cultures that pesticides such as DDT, chlordecone and toxaphene stimulated estrogen receptors endogenously expressed in MCF-7 cells [64]. Interestingly, not all agents stimulate the receptor, some function as antagonists. More intriguing is that some agents are an antagonist at one subtype but an agonist at the other. It appears that the ERα is most sensitive to the actions of a variety of compounds. Pesticide exposure in mammary cell tumors that express both ERα and ERβ has been shown to most effectively reduce the mRNA for ERα alone – with virtually no effect on ERβ mRNA expression (with the exception of chlorpyrifos). These effects were reversed when estradiol was added to the incubation media [65]. This report was further advanced by identifying prochloraz as a potential candidate for estrogenic-like effects. In a manner similar to β-estradiol, prochloraz downregulated the expression of both ERα and ERβ mRNA expression as well as reduce the expression of the ERα protein itself in the MCF-7 cell line [66]. An advancement used for the screening of compounds against actions at estrogen receptors is the use of reporter cell lines. These cells express both ERα and ERβ mRNA, and express function ERα and ERβ receptor protein. Kojima et al. [67] report the use of Chinese hamster ovary (CHO) cells which have been transfected with both ERα and ERβ cDNA. They screened nearly 200 pesticides – and nearly 25% demonstrated activity at ERα receptors, with nearly 15% exerting activity on the ERβ receptor [67]. Interestingly, they screened these compounds against the expression of androgen receptors (AR) – and their findings were that about 15 pesticides had both ER and AR activity, which would result in endocrine disruption in both males and females. Using HeLa cells with an estrogen reporter luciferase vector into two cell lines one each with the ERα or ERβ constructs permitted simultaneous screening of multiple compounds for agonistic or antagonistic activity [68]. Several pesticides (DDT, transnonachlor, chlordane, fenvalerate, and Toxaphene) were able to stimulate both ERα and ERβ [64, 68]. Only a few compounds demonstrated antagonistic properties at the ERα receptor (carbaryl, pentachlorophenol, 2, 4, 5-trichlorophenoxyacetic acid), yet significantly more pesticides were shown to be able to block the activity of the ERβ receptor (chlordecone, methoxychlor, carbaryl, endosulfan, endrin, dieldrin, and Aldrin) [68]. The negative effects of pesticides can be offset by interactions of the anticancer agents, 4-hydroxytamoxifen and trilostane. Both agents will block the estrogenic activity of compounds (estrogen and pesticides) at the estrogen receptor. In addition, these compounds will tend to downregulate protein kinase genes that were upregulated following exposure to pesticides [69]. Resistance may develop to the actions of 4-hydroxytamoxifen, which can be overcome with trilostane administration. In MCF-7 breast cancer cells, trilostane exposure was shown to increase the expression of ERβ [69]. Upregulation of ERβ may then provide additional ameliorative effects on the proliferation of breast cancer cells following the development of tolerance to first-line therapy. Collectively, the potency of these compounds was significantly less than that of β-estradiol, but continuous exposure to low levels of these compounds, or multiple compounds, may result in yet unknown consequences.

In addition to screening for estrogen receptor activity, other methodologies are being developed to further understand the actions of these endocrine disruptors. Estradiol hydroxylase activity has been examined by numerous investigators as a potential predictor of carcinogenic activity. In particular, estradiol 2-hydroxylase has received much attention as a potential predictor. Yet, the results have not been clear and the interpretation of these findings has been complicated, leading to the conclusion that estradiol 2-hyroxylase is not the best predictor of carcinogenic activity [70]. The enzyme "aromatase" is vital for the conversion of testosterone to β-estradiol. A logical extension would be to examine pesticides for their ability to inhibit aromatase (or CYP19 aromatase) activity similar to estrogen [71]. Under physiological conditions, estrogen would negatively feedback onto the aromatase enzyme reducing activity and reducing the amount of estrogen being synthesized from testosterone. Similar to the results observed in the estrogen receptor assays, fungicides such as prochloraz and imazalil inhibited aromatase activity to a greater extent than 4-hydroxyandrostendione. Other pesticides did inhibit aromatase activity but at significantly reduced efficacy and potency. Nearly 33% of the compounds tested exhibited some form of aromatase inhibition [72]. More recently, these findings were substantiated by Sanderson et al. [73]. They report that many of the compounds which may exhibit weak aromatase inhibition did so at concentrations that were cytotoxic to their cell system, R295R cells [73]. They did describe similar effects with fungicides and their effect on inhibiting aromatase activity, but it was suggested that these effects may not be through direct inhibition of aromatase activity, but through inhibition of phosphodiesterase activity [73]. Changes in human CYP19 aromatase expression may then lead to a predisposition to cancer development. One study examined a polymorphism in the CYP19 gene in a Greek population and linked this polymorphism, the population exposure to pesticides, and the incidence of breast cancer [74]. This study did not find a strong association between the short tandem repeat polymorphism, pesticide exposure, and breast cancer development. It is obvious that there are many avenues that can be traveled for the development of breast cancer and that many of the compounds that were viewed to be promoters of tumor formation may in fact function through multiple pathways, and with mixtures of agents, the potential interactions may go up exponentially. A recent report by Sitgaard-Kjeldsen et al. [75] substantiated these conclusions. Using mixtures of pesticides, they report that the observed outcomes are mediated by alterations in ERα, ERβ, androgen receptors, and aromatase [75]. Where understanding the effects of single compounds is important, more fully elucidating the combined actions of pesticides on the development of breast cancer is tantamount to completely understanding the actions of these compounds.

3. Heavy Metals/metalloestrogens in the Environment

Many heavy metals are naturally occurring and are used in a variety of industrial settings. Metals cannot be created or destroyed, but can change form, altering their biological availability and toxicity. Metals used in industry many times wind up in the food supply, groundwater, drinking water, and soil. Metals are found in many consumer products as well. These metals include copper, cobalt, nickel, lead, mercury, tin, chromium, cadmium, aluminum, vanadate (metal anion), antimony, barium, selenite, and arsenite. Cadmium is found in many farm fertilizers and can make its way into soil and water. Some of the other main sources of cadmium include cigarette smoke, rechargeable batteries, certain cosmetics, bread and other cereals, potatoes, root crops, and vegetables. Cadmium is a widely distributed metal used in manufacturing and present in a number of consumer products. It is used as a metal alloy, in paint, batteries (Ni-Cd), pigments, metal coatings, plastics, welding, and battery manufacture. Numerous heavy metals have been implicated in the development of a variety of cancers (arsenic, beryllium, cadmium, nickel, and hexavalent chromium, plus others). There have been an increasing number of studies which have shown that cadmium may facilitate breast cancer development. Nearly two decades ago, the presence of heavy metals as a potentially toxic addition to inorganic pesticide mixtures was addressed [76]. They report that multiple metals such as cadmium, cobalt, copper, and zinc can be found in relatively high levels as impurities in pesticide preparations. In other preparations, higher levels of lead, nickel, iron, and manganese were observed [76].

Metalloestrogens are a class of inorganic xenoestrogens which can affect the gene expression of human cells responding to estrogen. There have been numerous reviews on the subject of heavy metals as potential endocrine disruptors or metalloestrogens [77-80]. The most extensively studied of all of the metalloestrogens is cadmium. Cadmium is found in the air (ambient, occupational, and cigarette smoke). Under normal conditions, the concentration of cadmium in the air is relatively low. Occupational exposure limits to cadmium have been lowered 50-100-fold in the last few decades as the toxic effects of cadmium were better understood. Cadmium can also been found in the soil and in groundwater. Examining sources of cadmium contamination, the greatest source of human exposure is phosphate fertilizers, followed by fossil fuel combustion (automobile exhaust), iron and steel production, natural source, cement production, cadmium-containing products, and finally incineration. Over ½ of all of the cadmium exposure is through exposure to phosphate fertilizer and automobile exhaust. Cadmium has been found in higher concentrations in males compared to females and can affect steroid function in both genders by interfering with the biosynthesis of androgens, estrogens, and progesterone [80]. The majority of the published literature has focused on estrogen effects versus androgen effects. Dyer [79] and Byrne et al. [81] did review some of the published androgenic effects of heavy metals on testicular function. The effects of metalloestrogens are related to the physiologic function of estrogen because metalloestrogens have shown affinity for estrogen receptors. Because they can mimic estrogen, thus activating the receptor, they are considered harmful and potentially linked with breast cancer [7]. Heavy metals including copper, cobalt, nickel, lead, mercury, tin, chromium, and vanadate exhibited estrogenic activity when subjected to a variety of tests, resulting in a two- to fivefold increase in the number of human breast cancer cells [82]. In another study, two different screening systems were used to test for "estrogenicity" in heavy metals, ultimately verifying the estrogen-like activity of these metals. Antimony and barium were also implicated as being estrogenic [8]. In the Choe et al. [8] study, the potency of cadmium in eliciting estrogenic effects was second only to n-butyltin. In vivo studies which utilized a concentration of cadmium that could be considered "environmentally-relevant, " elicited numerous changes in female rats [83]. Changes were the most pronounced in estrogen-dependent tissues such as the mammillary glands. In general, there was an observed upregulation in the development of the glands, increased milk production, increased receptor density (estrogen and progesterone). Collectively, all of these changes could lead to the overdevelopment of breast tissues, and the development of particular cancers. Part of the effects elicited by cadmium resembles those of effects attributed to estrogen, which has led to cadmium being considered a "metalloestrogen". Similar to the effects of pesticides on fetal development, the exposure of the female fetus to cadmium can also result in the accelerated growth and development of breast tissue [83]. In sum, all of these changes have been associated with an increased risk for developing breast cancer.

3.1. Metalloestrogen link to breast cancer

The exposure to metalloestrogen metals has increased dramatically over the last half century. This has occurred through an increase in environmental release of these metals and an increased contamination with the food and water supply and through smoking of tobacco products. It has been suggested that these agents can stimulate estrogen receptors in the absence of estrogen, leading to a potentially negative health impact in postmenopausal women. Further compounding their health hazards is the extremely long biological half-life (cadmium is 10 to 30 years) and the ability to bioaccumulate [81, 84]. Cadmium and other heavy metals have been shown to be estrogenic, and have the ability to activate the estrogen receptor, similar to the effects of estradiol [82, 85]. In vitro assays examining estrogen responsiveness using various metals demonstrated potencies of 25-100% of estradiol [82, 86]. In vivo assays demonstrated that acute administration of metals (including cadmium) elicited classic estrogenic responses with an increase in uterine weight, hyperplasia, and hypertrophy of the endometrial lining, increased expression and density of progesterone receptors and increase in mammary tissue density [82, 86]. A recent study focusing solely on mammary developing following endocrine disruptor administration showed that cadmium can alter normal mammary tissue development, gene expression, and that early exposure may lead to a predisposition toward breast cancer development [21]. Cadmium can mimic the action of estrogen and appears to recognize the estrogen binding site on both the ERα and ERβ receptors. Some of this action is via the estrogen receptor mediated by the GPR30 receptor [85]. Via the GPR30 receptor, cadmium stimulation can activate MAPK, raf-1, and ERK1/2 [85]. The expression of estrogen receptors is also affected and results in the levels of estrogen receptor being greatly reduced [87], again mimicking the effects of natural estrogens. These, and other, in vitro assays have shown that cadmium is estrogenic and can mediate estrogen receptor action, as well as the mitogenic effects of estrogen receptor stimulation. Yet, there have been mixed results in other experimental systems suggesting more work needs to be performed [88, 89]. The regulation of expression and activity of estrogen receptor plays an essential role in the growth, differentiation, and prognosis of human breast cancer. Thus, the effects of the metals in living organisms have also been found to change the breast anatomy in specific ways, making it more susceptible to cancer, including an earlier onset of puberty, an increase in epithelial area, and an increase in the number of terminal end buds in the mammary gland [83]. Aluminum has been found to interact with DNA, binding strongly to the phosphate backbone of the structure under neutral conditions [90]. This suggests that aluminum could serve as a possible source of DNA damage, increasing the chances of DNA mistakes and promoting the growth of the damaged cells. It has also been shown that aluminum can interfere with cell growth regulatory processes through many pathways, including altering gene expression [77]. The correlation between breast cancer and heavy metals is not confined to a few metals, but numerous metals have been identified in breast biopsies in significant concentrations. This is a correlation not observed when compared to biopsies from healthy breast tissue [91, 92]. Similar to what has been reported regarding the correlation of pesticide exposure to breast cancer, there are some questions regarding the function of cadmium as a metalloestrogen which can promote breast cancer proliferation [93]. In the majority of in vitro assays, the data are fairly convincing regarding the metalloestrogen effects of cadmium. The actions of cadmium have provided the more compelling evidence of a correlation with breast cancer proliferation [84]. The affinity of cadmium for the estrogen receptor (ERα) is comparable to that for estrogen (0.4-0.5 nM), yet the actual binding site of cadmium to the estrogen receptor is yet to be determined [84]. One current belief is that ERα is sequestered in the inactive form – low estrogen or no estrogen. Once activated, the receptor undergoes a conformational change and enters the active state. Increasing the number of active ERα receptors will increase estrogen responsiveness [81]. Cadmium can block the action of estrogen suggestion that at least parts of its effects are mediated through the estrogen binding site on ERα. Molecular studies have further supported this hypothesis examining various mutations and the ability of a selective estrogen ERα antagonists to block estrogen effects in transfected COS-7 cells through interactions at the hormone binding domain on ERα [94]. Through mutation studies, potential interaction sites have been identified as cys381, cys447, glu523, his524, and asp538 [81, 94]. The actions at estrogen receptors, the activation of various protein kinases, and the increased proliferation of breast cancer cells have been replicated by a variety of investigators in a variety of assay systems. In humans, these findings are not quite as clear. Whereas some studies have suggested a correlation between cadmium content and breast cancer [91, 92], Silva et al. [93] review numerous studies where the data are inconclusive. As with pesticide exposure, the discovery of biomarkers specific for cadmium exposure is necessary. Although the data are not definitive for being a causative agent, cadmium can be considered a risk factor for breast cancer development. Another drawback to interpreting in vitro data is that these studies usually involve acute exposure. In humans, exposure to cadmium results in years if not decades of low-level exposure that results in an increasing bioaccumulation and increased cadmium burden on the patient. Acute in vitro studies can examine receptor and nonreceptor-mediated intracellular effects in the relative short term. Real-world exposure would involve a very gradual change in intracellular signaling systems, leading to changes that may lead to indirect changes that are not being observed in the current in vitro experimental paradigm.

4. Combination of Metalloestrogens and Pesticides

There has been extensive work done examining the individual actions of heavy metals or pesticides in the environment. There has been virtually no work done examining the combined effects of heavy metals and pesticides in the environment. This relative dearth of information has left a significant void in our understanding of the actions of these compounds and their potential role in the development of breast cancer. Many of the compounds discussed in this chapter have demonstrated the ability to interact with many of the pathways associated with tumorigenesis. Interactions with the function of a variety of caspases, Akt/mTOR pathway, p53, ERK1 and 2 signaling pathways, etc. have created a clear need for extensive work in these areas. In many instances, there are no clear direct interactions, but may in fact act through various steps in one or more of these cell cycle cascades. It is quite clear that pesticides and heavy metals do coexist in the environment. There have been numerous reports outlining the effects/coexistence of these compounds in various ecosystems and at various levels of the food chain.

Current knowledge is lacking regarding the collective effects of estrogen and all of the various estrogen-like compounds. Only in the last two decades has there been increasing amounts of work examining the estrogenic effects of naturally occurring compounds as well as synthetic compounds such as pesticides and metals. What is unknown is what overall effect all of these agents will have on a person. These effects have been shown as early as following in utero exposure [95] or through prolonged exposure to trace levels of contaminants leading to endometriosis [96]. The basic physiological responses for estrogen are to control secondary sexual characteristics, influence metabolic activity, central nervous system effects, effects on bone turnover, and, through aromatase, interplay with testosterone. Estrogen can exert these effects by a receptor-mediated mechanism – through estrogen receptor alpha [ERα], estrogen receptor beta [ERβ], and through GPR30. Once stimulated, these receptors will work through intracellular mechanisms leading to nongenomic and genomic responses, and with improving technology, we know that estrogen can regulate hundreds of genes – with the majority (~70%) being downregulated [97, 98]. Since complete pathways have not been elucidated for each of the main classes of exogenous estrogenic compound, it is not known or it is unclear whether there are additive, synergistic, or potentiating effects of these compounds when humans are coexposed. This is one of the troubling aspects of our current knowledge. We have little idea of what effects polyexposure will have on a human, and at what level or threshold will we begin to see these effects. An enormous amount of work still needs to be done in these areas to determine the safe exposure levels, not just for the known compounds, but also for combinations of the compounds [99, 100].

Not only are the potential pathways taken by each of the estrogenic compounds highly complex, but the process of carcinogenesis is also very complex. These processes can be highly involved with many steps and processes working in concert to finally yield a malignant transformation [11]. Cancers may be single-cell in origin and with the mutation of a few genes, errors are expressed that lead to errors in replication and/or growth. In addition, these malignant/mutated cells, dependent on the environment (pollution, toxins, etc.) can cause the conversion of otherwise "normal" cells to cancerous [11]. These "gene–environment interactions" (GEI) are broadly defined as interactions between environmental exposures and specific (risk) genotypes. The term GEI refers to the joint influences of genetic and environmental factors on health and disease. Environmental exposures affect gene regulation and/or act as additive risk factors in conjunction with a particular allelic form of a gene (genetic polymorphism), influencing disease initiation and progression. GEI also entails the different effects of a given environmental exposure on individuals and the different effects of a genotype in people with different histories of environmental exposure. For an excellent review of GEI and the mechanism(s) by which GEI can lead to cancer formation, refer to Tabrez et al. [11]. Kiyama et al. [98] reported a comprehensive listing of various genes which are regulated by estrogen and reviewed their activity. They focused on the classical "endocrine disruptors, " by focusing on their cell signaling. Their studies first examine gene expression profiles, followed by cell signaling responses. The signaling pathways identified could be used as candidate toxicity pathways to monitor and evaluate endocrine disruptor action [98].

In addition to in vitro studies, there have been human studies which have addressed the concerns associated with the combinations of pesticides and heavy metals. There is a considerable burden of in vitro and in vivo immunotoxicity evidence regarding the detrimental actions of pesticides and heavy metals. Yet, as evidence and information mounts, the findings are still far from unequivocal. Between differences in study design, test subjects, data analysis, and model systems used, etc., it has been virtually impossible to develop a clear correlation between these environmental agents and incidence of disease [99, 100]. Making the correlation between low-level exposures in animal studies to the immune system altering effects observed in humans has been difficult. Also, the effects on human health of the synergistic interactions between natural, medical, dietary, and environmental estrogens have not been fully elucidated yet. There are several factors which need to be accounted for when examining the effects of environmental estrogens: 1) immune status (immunocompromised would have larger response) of the individual, 2) gender (females more responsive than males), 3) status of poly-pollutant exposure, and 4) duration of the exposure. Exposure to the metalloestrogen arsenite in utero altered mammary gland development prior puberty [95]. There was an overgrowth leading up to puberty and a densifying of the breast tissue. After puberty, there was a clear upregulation in the density of estrogen receptor-alpha (ERα) due to the increased and altered response of the ERα transcripts [95], which may lead to the increased risk of developing breast cancer. The large "ENDO Study" examined 22 trace elements and found that 19 of the 22 (86%) of the elements (mostly heavy metals) that were examined were not correlated with endometriosis [96]. Yet, the remaining three that did appear to correlate with endometriosis were cadmium, chromium, and copper – 3 metals that are known metalloestrogens. Additional work will need to be done to substantiate these findings and support their conclusions.

5. Conclusions

Over the past 20-30 years, it has become increasingly evident that pesticide use can have unwanted physiological effects beyond the acute exposure. Many pesticides have been banned in numerous countries as these health effects have been described. Even with banning some pesticides, there are stockpiles that need to be disposed of, and some are still used frequently in developing nations. Initial toxicology analysis has focused on the toxicity of individual compounds but this method of assessment may significantly underestimate the risk associated with these compounds when found in mixtures. Many health organizations are now calling for retesting of these agents, but with new guidelines for assessing the potential risk to human and animal life [101, 102]. Through a variety of complex mechanisms, many of these agents have been shown to interact at the estrogen receptor, both ERα and ERβ. These effects can occur in the absence of estrogen and can potentiate estrogenic effects in many mammalian tissues, such as breast and uterus. These interactions lead to the hypothesis that particular pesticide agents – such as organochlorine pesticides – may disrupt the natural endocrine function (i.e., "endocrine disruptors") of the organism. Interference with the reproductive systems of aquatic- and land-based wildlife may lead to dwindling populations of species of fish, shellfish, and mammals, potentially leading to their extinction. In humans, studies to establish the correlation between pesticide exposure and breast cancer has not been clear and absolute. There has been evidence of positive correlation within some ethnic populations, whereas other studies have yielded negative correlations. In most instances, these were retrospective studies and the design, subject inclusion, and the number of subjects has limited the ability to draw strong conclusions. Obviously, the effects of long-term human exposure needs further study with strict guidelines. There also needs to be a strengthening of the toxicity testing of pesticide mixtures to avoid underestimating the potential toxic effects.

Select heavy metals have been shown to have estrogenic properties and have since been referred to as "metalloestrogens." Of these, cadmium has been the most extensively studied and appears to be the most potent metalloestrogen at stimulating the estrogen receptor. Both the affinity and inhibitory constant at the ERα receptor is approximately 0.5 nM, which is in order with the affinity of estrogen for its receptor. In vitro studies have shown that cadmium, and some other metalloestrogens, can elicit estrogenic effects resulting in elevation of both ERα and ERβ receptor densities, increased expression and activity of intracellular protein kinases, increased density of progesterone receptors, and the increased size of the uterus as well as increased development and proliferation of breast tissues. Collectively, all of these responses in the presence of metalloestrogens led to speculation that metalloestrogens may be correlated to the incidence of breast cancer. Most of the data currently available have been anecdotal, and have involved in vitro assay systems and breast cancer cell lines (such as MCF-7). In these systems, it is clear that cadmium is the most potent of the metalloestrogens at stimulating tumor development through direct actions at the estrogen receptor, as well as intracellular effects which may be indirect but involve the activation of many signaling systems implicated in tumor development. In human studies, these correlations are not as clear. In a few studies, there was a higher concentration of cadmium in the breast tissue compared to controls, but the direct relationship with breast cancer progression was not clear. Other studies have shown no relationship between cadmium exposure and breast cancer development. One shortcoming of the in vitro studies is that they are relatively acute exposure, for a short duration. This makes it extremely difficult to draw comparisons to human exposure. The potential that low-level, decade-long exposure to cadmium (remembering that cadmium will bioaccumulate, so exposure could be just from the body burden) may result in subtle changes which increase the predisposition to breast cancer development. Many groups are now calling for additional investigation into the mechanisms by which metalloestrogens exert their effects. This additional work is critical to better understanding the actions of metalloestrogens at the estrogen receptors (through NMR or X-ray crystallography), and the interaction of metalloestrogens on intracellular signaling systems.

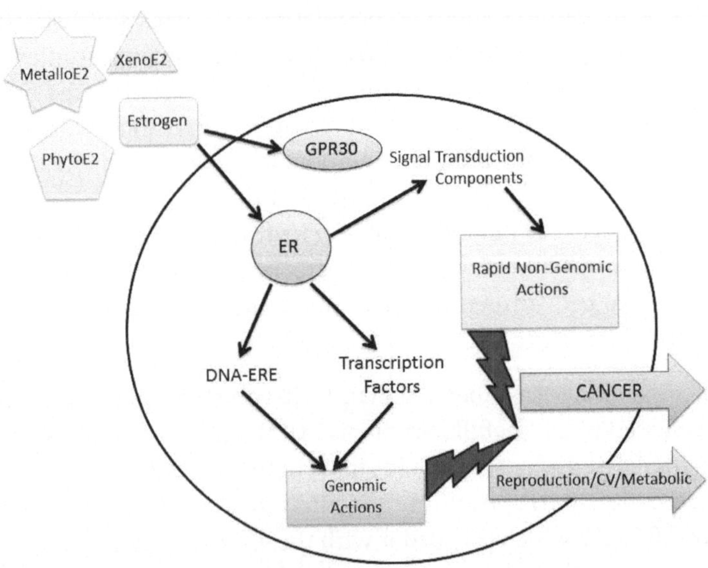

Figure 2. General schematic of the effects of estrogen, xenoestrogens [XenoE2], metalloestrogens [MetalloE2], and phytoestrogens [PhytoE2] on cellular function and the resulting physiological/pathological outcomes. Modified from Darbe, 2014 [97] [PhytoE2 = isoflavones, etc.; XenoE2 = pesticides, etc.; MetalloE2 = heavy metals like cadmium, etc.].

Lastly, to believe that an individual would be exposed to only one agent would be naïve. For example, a smoker working in the pesticide industry would undoubtedly have elevated cadmium levels due to the tobacco smoke, and possible passive exposure to the pesticides through dermal absorption or inhalation. The combination of these compounds may have a great additive effect on the development of breast cancer. Currently, there are few studies that address these types of combination exposures and virtually none involving the human population. Collectively, as our understanding has grown regarding the negative health effects of pesticides and metalloestrogens, a significantly greater number of questions have arisen and has shed light onto obvious gaps in our understand. As much work that has been done in the last 30 years or so, even more work needs to be done in the next decade or two to assist in our understanding of the pathogenesis of diseases like breast cancer. By better understanding the root causes and foundations for breast cancer development, we will be able to focus our toxicological investigations onto those causes. Also, as we improve our understanding, both in the development of breast cancer, and the involvement of pesticides and metalloestrogens, we will be able to develop better biomarkers. It may not be impossible to completely eliminate exposure, but a viable biomarker that will predict with a high degree of reliability will help with therapeutic interventions at a much earlier time, thereby reducing the morbidity and mortality of this disease.

References

1. Rao GVR, Rupela OP, Rao VR, et al. Role of biopesticides in crop protection: present status and future prospects. Ind J Plant Protect 2007; 35(1):1–9.
2. Ritter SK. Pinpointing Trends In Pesticide Use. Chemical and Engineering News 2009; 87(7) http://cen.acs.org/articles/87/i7/Pinpointing-Trends-Pesticide-Use.html (accessed February 2, 2015)
3. Pimentel D. Environmental and economic costs of the application of pesticides primarily in the United States. Environment, Development and Sustainability 2005; 7:229-252 DOI 10.1007/s10668-005-7314-2
4. Pimentel D, Acquay H, Biltonen M, et al. Environmental and economic costs of pesticide use. BioScience 1992; 42:750-760
5. Cooper J, Dobson H The benefits of pesticides to mankind and the environment. Crop Protect 2007; 26: 1337–1348 DOI 10.1016/j.cropro.2007.03.022
6. Breast Cancer Fund. Clear Science: Chemicals in Household Products. www.breastcancerfund.org/clear-science (accessed 20 January 2015)
7. Darbre PD Environmental oestrogens, cosmetics and breast cancer. Best Prac Res Clin Endocrinol 2006; 20:121-143
8. Choe SY, Kim SJ, Kim HG, et al. Evaluation of estrogenicity of major heavy metals. Sci Total Environ. 2003; 312:15-21 DOI 10.1016/S0048-9697(03)00190-6
9. Kakkar P, Jaffery FN. Biological markers for metal toxicity. Environ Toxicol Pharmacol 2005; 19:335-349

10. Helmfrid I, Berglund M, Lofman O, et al. Health effects and exposure to polychlorinated biphenyls (PCBs) and metals in a contaminated community. Environ Int 2012, 44:53-58

11. Tabrez S, Priyadarshini M, Priyamvada S, et al. Gene-environment interactions in heavy metal and pesticide carcinogenesis, Mutation Research/Genetic Toxicology and Environmental Mutagenesis 2014; 760:1-9 DOI 10.1016/j.mrgentox.2013.11.002

12. Aronson KJ, Miller AB, Woolcott CG, et al. Breast adipose tissue concentrations of polychlorinated biphenyls and other organochlorines and breast cancer risk. Can Epidemiol Biomarkers & Prevention 2000; 9:55-63 (Downloaded from cebp.aacrjournals.org on February 3, 2015)

13. Safe S. Endocrine disruptors and human health: is there a problem. Toxicology 2004; 205(1–2):3–10 DOI 10.1016/j.tox.2004.06.032

14. Murray DW, Lichter SR. Organochlorine residues and breast cancer. New Eng J Med 1998; 338(14):990–991 DOI 10.1056/nejm199804023381411

15. Calle EE, Frumkin H, Henley SJ, et al. Organochlorines and breast. Cancer Risk CA Can J Clinic 2002; 52:301-309 DOI 10.3322/canjclin.52.5.301

16. Caserta D, Maranghi L, Mantovani A, et al. Impact of endocrine disruptor chemicals in gynecology. Human Reproduction Update 2008; 14(1):59–72 DOI 10.1093/humupd/dmm025

17. Roy JR, Chakraborty S, Chakraborty TR. Estrogen-like endocrine disrupting chemicals affecting puberty in humans – a review. Med Sci Mon 2009; 15(6):RA137-145

18. Patisaul HB, Adewale HB Long-term effects of environmental endocrine disruptors on reproductive physiology and behavior. Front Behavior Neurosci 2009; 3(10):1-18 DOI 10.3389/neuro.08.010.2009

19. Karalliedde L, Henry J The acute cholinergic syndrome. In: Organophosphates and Health (Karalliedde L, Feldman S, Henry J, Marrs T, eds) 2001; River Edge, World Scientific Publishing ISBN: 978-1-78326-143-7

20. Environmental Protection Agency [EPA-HQ-OPPT-2004-0109; FRL-8399-7] Final List of Pesticide Active Ingredients and Pesticide Inert Ingredients to be Screened Under the Federal Food, Drug and Cosmetic Act 2009; 74(71):17579-17585

21. Osborne G, Rudel R, Schwarzman M. Evaluating chemical effects on mammary gland development: A critical need in disease prevention. Reprod Toxicol 2014, http://dx.doi.org/10.1016/j.reprotox.2014.07.077

22. Clarke BO, Porter NA, Marriott PJ et al. Investigating the levels and trends of organochlorine pesticides and polychlorinated biphenyl in sewage sludge. Environ Int; 2010; 38:323-329

23. Schaefer WR, Hermann T, Meinhold-Heerlein I et al. Exposure of human endometrium to environmental estrogens, antiandrogens and organochlorine compounds. Fert Steril 2000; 74:558-563

24. Saoudi A, Frery N, Zeghnoun A, et al. Serum levels of organochlorine pesticides in the French adult population: The French National Nutritional and Health Study (ENNS), 2006-2007. Sci Total Environ 2014; 472:1089-1099
25. Konar SK, Mullick M. Problems of safe disposal of petroleum products, detergents, heavy metals and pesticides to protect aquatic life. Sci Total Environ 1993; Supplement:989-1000
26. Becker PR. Concentration of chlorinated hydrocarbons and heavy metals in Alaska arctic marine mammals. Mar Poll Bull 2000; 40(10):819-829
27. Kumar KS, Sajwan KS, Richardson JP, et al. Contamination profiles of heavy metals, organochlorine pesticides, polycyclic aromatic hydrocarbons and alkylphenols in sediment and oyster collected from marsh/estuarine Savannah GA, USA. Mar Poll Bull 2008; 56:136–162
28. Maes J, Belpaire C, Goemans G. Spatial variations and temporal trends between 1994 and 2005 in polychlorinated biphenyls, organochlorine pesticides and heavy metals in European eel (Anguilla anguilla L.) in Flanders, Belgium. Environ Poll 2008; 153:223-237 DOI 10.1016/j.envpol.2007.07.021
29. Yatawara M, Qi S, Owago OJ et al. Organochlorine pesticide and heavy metal residues in some edible biota collected from Quanzhou Bay and Xinghua Bay, Southeast China. J Environ Sci 2010; 22(2):314–320 DOI 10.1016/S1001-0742(09)60110-8
30. Mazet A, Keck G, Berny P. Concentrations of PCBs, organochlorine pesticides and heavy metals (lead, cadmium, and copper) in fish from the Drôme River: Potential effects on otters (Lutra lutra). Chemosphere 2005; 61:810-816 DOI 10.1016/j.chemosphere.2005.04.056
31. Sankar TV, Zynudheen AA. Anandan R, et al. Distribution of organochlorine pesticides and heavy metal residues in fish and shellfish from Calicut region, Kerala, India. Chemosphere 2006; 65:583-590 DOI 10.1016/j.chemosphere.2006.02.038
32. Araújo DFS, Silva AMRB, Lima LLA The concentration of minerals and physicochemical contaminants in conventional and organic vegetables. Food Contr 2014; 44:242-248 DOI 10.1016/j.foodcont.2014.04.005
33. Mansour SA, Belal MH, Abou-Arab AAK, et al. Monitoring of pesticides and heavy metals in cucumber fruits produced from different farming systems. Chemosphere 2009; 75:601–609 DOI 10.1016/j.chemosphere.2009.01.058
34. Nasreddine Parent-Massin L. Food contamination by metals and pesticides in the European Union. Should we worry? Toxicol Lett 2002; 127:29-41
35. Mansour SA, Gad MF, Risk assessment of pesticides and heavy metals contaminants in vegetables: A novel bioassay method using Daphnia magna Straus. Food Chem Toxicol 2010; 48:377–389 DOI 10.1016/j.fct.2009.10.026
36. Li B, Wu S, Lui C. China pharmaceuticals: Investing in Traditional Chinese Medicine (TCM). Hong Kong, China: Morgan Stanley Research Asia/Pacific; 2009.

37. Nahin RL, Barnes PM, Stussman BJ et al. Costs of complementary and alternative medicine (CAM) and frequency of visits to CAM practitioners: United States, 2007. National Health Status Report 2009; 18:1–15. http://www.cdc.gov/nchs/data/nhsr/nhsr018.pdf (accessed February 6, 2015)
38. Varma SP, Boldin BR, Lin PS. The Inhibition of the estrogenic effects of pesticides and environmental chemicals by curcumin and isoflavonoids. Environ Health Persp 1998; 106(12):807812
39. Ernst E. Toxic heavy metals and undeclared drugs in Asian herbal medicines. Trends Pharmacol Sci 2002; 23:136–139 DOI 10.1016/S0165-6147(00)01972-6
40. Ernst E, Coon JT. Heavy metals in traditional Chinese medicines: A systematic review. Clin Pharmacol Therapeut 2001; 70:497–504 DOI 10.1067/mcp.2001.120249
41. Lin CG, Schaider LA, Brabander DJ et al. Pediatric lead exposure from imported Indian spices and cultural powders. Pediatrics 2010; 125:E828–35 DOI 10.1542/peds.2009-1396
42. Harris ESJ, Cao S, Littlefield BA, et al. Heavy metal and pesticide content in commonly prescribed individual raw Chinese herbal medicines. Sci Total Environ 2011; 409:4297–4305 DOI 10.1016/j.scitotenv.2011.07.032
43. Liu J, Lu YF, Wu Q et al. Mineral arsenicals in traditional medicines: Orpiment, realgar, and arsenolite. J Pharmacol Exper Therapeut 2008a; 326:363–368 DOI 10.1124/jpet.108.139543
44. Liu J, Shi JZ, Yu LM, et al. Mercury in traditional medicines: is cinnabar toxicologically similar to common mercurials? Exper Biol Med 2008b; 233:810–817 DOI 10.3181/0712-MR-336
45. Saper RB, Phillips RS, Sehgal A, et al. Lead, mercury, and arsenic in US- and Indian-manufactured Ayurvedic medicines sold via the Internet. J Am Med Assoc 2008; 300:915–923 DOI 10.1001/jama.300.8.915
46. US GAO (United States Government Accountability Office). Herbal dietary supplements: examples of deceptive or questionable marketing practices and potentially dangerous advice. Washington, DC, United States: US GAO; 2010; GAO-10-662T http://www.gao.gov/new.items/d10662t.pdf (accessed February 6, 2015)
47. Dich J, Zahm SH, Hanberg A, Adami HO. Pesticides and cancer. Can Causes Cont 1997; 8:420-443
48. George J, Shukla Y. Pesticides and cancer: Insights into the toxicoproteomic-based findings. J Proteomics 2011; 74:2719-2722 DOI 10.1016/j.prot.2011.09.024
49. Mostafalou S, Abdollahi M. Pesticides and human chronic diseases: Evidences, mechanism, and perspectives. Toxicol Appl Pharmacol 2013; 268:157-177 DOI 10.1016/j.taap.2013.01.025
50. Snedeker SM. Pesticides and breast cancer risk: A review of DDT, DDE and Dieldrin. Environ Health Persp 2001; 109(1):35-47

51. Ingber SZ, Buser MC, Pohl HR, et al. DDT/DDE and breast cancer: A meta-analysis. Regul Toxicol Pharmacol 2013; 67:421-433 DOI 10.1016/j.yrtph.2013.08.021
52. Mathur V, Bhatnagar P, Sharma RG, et al. Breast cancer incidence and exposure to pesticides among women originating from Jaipur. Environ Int 2002; 28:331-336
53. Landau-Ossondo M, Rabia N, Jos-Pelage J, et al. Why pesticides could be a common cause of prostate and breast cancers in the French Caribbean Island, Martinique. An overview on key mechanisms of pesticide-induced cancer. Biomed Pharmacother 2009; 63:383-395 DIO 10.1016/j.biopha.2009.04.043
54. O'Leary ES, Vena JE, Freudenheim JL, et al. Pesticide exposure and risk of breast cancer: a nested case-control study of residentially stable women living on Long Island. Environ Res 2004; 94:134-144 DOI 10.1016/l.envres.2003.08.001
55. Reynolds P, Hurley SE, Goldberg DE, et al. Residential proximity to agricultural pesticide use and the incidence of breast cancer in the California Teachers Study cohort. Environ Res 2004; 96:206-218 DOI 10.1016/j.envres.2004.03.001
56. Bradlow HL, Davis D, Sepkovic DW, et al. Role of the estrogen receptor in the action of organochlorine pesticides on estrogen metabolism in human breast cancer cell lines. Sci Total Environ 1997; 208:9-14
57. Hodges LC, Bergerson JS, Hunter DS et al. Estrogenic effects of organochlorine pesticides on uterine leiomyoma cells in vitro. Toxicol Sci 2000; 54:355-364
58. Ventura C, Nunez M, Miret N, et al. Differential mechanisms of action are involved in chlorpyrifos effects in estrogen-dependent or –independent breast cancer cells exposed to low or high concentrations of the pesticide. Toxicol Lett 2012; 213:184-193 DOI 10.1016/j.toxlet.2012.06.017
59. Ventura C, Venturino A, Miret N, et al. Chlorpyrifos inhibits cell proliferation through ERK1/2 phosphorylation in breast cancer cell lines. Chemosphere 2015; 120:343-350 DOI 10.1016/j.chemosphere.2014.07.088
60. `Aube M, Larochelle C, Ayotte P. Differential effects of a complex organochlorine mixture on the proliferation of breast cancer cell lines. Environ Res 2011; 111:334-347 DOI 10.1016/j.envres.2011.01.010
61. Valeron PF, Pestano JJ, Luzardo OP, et al. Differential effects exerted on human mammary epithelial cells by environmentally relevant organochlorine pesticides either individually or in combination. Chem-Biol Interact 2009; 180:485-491 DOI 10.1016/j.cbi.2009.04.010
62. Li X, Zhang S, Safe S. Activation of kinase pathways in MCF-7 cells by 17β-estradiol and structurally diverse estrogenic compounds. J Steroid Biochem Mol Biol 2006; 98:122-132 DOI 10.1016/j.jsbmb.2005.08.018
63. Vom Saal FS, Nagel SC, Palanza P et al. Estrogenic pesticides: binding relative to estradiol in MCF-7 cells and effects of exposure during fetal life on subsequent territorial behavior in male mice. Toxicol Lett 1995; 77:343-350

64. Soto Am, Chung KL, Sonnenschein C. The pesticides endosulfan, Toxaphene, and dieldrin have estrogenic effects on human estrogen-sensitive cells. Human Health Persp 1994; 102:380-383

65. Grunfeld HT, Bonefeld-Jorgensen EC. Effect of in vitro estrogenic pesticides on human estrogen receptor α and β mRNA levels. Toxicol Lett 2004; 151:467-480 DOI: 10.1016/toxlet.2004.03.021

66. Hofmeister MV, Bonefeld-Jorgensen EC. Effects of the pesticides prochloraz and methiocarb on human estrogen receptor α and β mRNA levels analyzed by online RT-PCR. Toxicol In Vitro 2004; 18:427-433 DOI10.1016/j.tiv.2003.12.008

67. Kojima H, Katsura E, Takeuchi S et al. Screening for estrogen and androgen receptor activities in 200 pesticides by in vitro reporter gene assays using Chinese hamster ovary cells. Environ Health Persp 2004; 112(5):524-531 DOI 10.1289/ehp.6649

68. Lemaire G, Mnif W, Mauvais P et al. Activation of α- and β-estrogen receptors by persistent pesticides in reporter cell lines. Life Sci 2006; 79:1160-1169 DOI 10.1016/j.lfs.2006.03.023

69. Barker S, Malouitre SDM, Glover HR, et al. Comparison of effects of 4-hydroxy tamoxifen and Trilostane on estrogen-regulated gene expression in MCF-7 cells: Up-regulation of estrogen receptor beta. J Steroid Biochem Mol Biol 2006; 100:141-151 DOI 10.1016/j.jsbmb.2006.04.006

70. McDougal A, Wilson C, Safe S. Induction of estradiol 2-hydroxylase activity in MCF-7 human breast cancer cells by pesticides and carcinogens. Environ Toxicol Pharmacol 1997; 3:195-199

71. Andersen HR, Vinggaard AM, Rasmussen TH, et al. Effects of currently used pesticides in assays for estrogenicity, androgenicity, and aromatase activity in vitro. Toxicol Appl Pharmacol 2002; 179:1-12 DOI 10.1006/taap.2001.9347

72. Vinggaard AM, Hnida C, Breinholt V et al. Screening of selected pesticides for inhibition of CYP19 aromatase activity in vitro. Toxicol In Vitro 2000; 14:227-234

73. Sanderson JT, Boerma J, Lansbergen GWA et al. Induction and inhibition of aromatase (CYP19) activity by various classes of pesticides in H295R human adrenocortical carcinoma cells. Toxicol Appl Pharmacol 2002; 182:44-54 DOI 10.1006/taap.2002.9420

74. DialynaI, Tzanakakis G, Dolapsakis G, et al. A tetranucleotide repeat polymorphism in the CYP19 gene and breast cancer susceptibility in a Greek population exposed and not exposed to pesticides. Toxicol Lett 2004; 151:267-271 DOI 10.1016/j.toxlet.2004.01.024

75. Stigaard-Kjeldsen L, Ghisari M, Bonefeld-Jorgensen EC. Currently used pesticides and their mixtures affect the function of sex hormone receptors and aromatase enzyme activity. Toxicol Appl Pharmacol 2013; 272:453-464 DOI 10.1016/j.taap.2013.06.028

76. Gimeno-Garcia E, Andreu V, Boluda R. Heavy metals incidence in the application of inorganic fertilizers and pesticides to rice farming soils. Environ Poll 1996; 92(1):19-25

77. Darbre PD. Review article: Underarm cosmetics and breast cancer. J Appl Toxicol 2003; 23:89-95

78. Darbre P. Metalloestrogens: An emerging class of inorganic xenoestrogens with potential to add to the estrogenic burden of the human breast. J Appl Toxicol 2005; 26:191-197 DOI: 10.1002/jat.1135

79. Dyer CA. Heavy metals as endocrine-disrupting chemicals. In: Endocrine-Disrupting Chemicals: From Basic Research to Clinical Practice (ed. Gore AC) 2007; Humana Press Inc. Totowa NJ, 111-133

80. Georgescu B, Georgescu C, Daraban S, et al. Heavy metals acting as endocrine disruptors. Anim Sci Biotechnol 2011; 44(2):89-93

81. Byrne C, Divekar SD, Storchan GB, et al. Cadmium – A metallohormone. Toxicol Appl Pharmacol 2009; 238:266-271 DOI: 10.1016/j.taap.2009.03.025

82. Martin MB, Reiter R, Pham T, et al. Estrogen-like activity of metals in MCF-7 breast cancer cells. Endocrinology 2003; 144:2425-2436

83. Johnson MD, Kenney N, Stoica A, et al. Cadmium mimics the in vivo effects of estrogen in the uterus and mammary gland. Natur Med 2003; 9:1081-1084

84. Aquino NB, Sevigny MB, Sabangan J, et al. The role of cadmium and nickel in estrogen receptor signaling and breast cancer: metalloestrogens or not? J Exper Sci Health Part C 2012; 30:189-224 DOI 10.1080/10590501.2012.705159

85. Byrne C, Divekar SD, Storchan GB et al. Metals and breast cancer. J Mamm Glan Biol Neoplas 2013; 18:63-73 DOI: 10.1007/s10911-013-9273-9

86. Yu X, Filardo EJ, Shaikh ZA. The membrane estrogen receptor GPR30 mediates cadmium-induced proliferation of breast cancer cells. Toxicol Appl Pharmacol 2010; 245:83–90 DOI 10.1016/j.taap.2010.02.005

87. Safe S. Cadmium's disguise dupes the estrogen receptor. Nature Medicine 2003; 9(8):1000-1001

88. Garcia-Morales P, Saceda M, Kenney N, et al. Effect of cadmium on estrogen-receptor levels and estrogen-induced responses in human breast cancer cells. J Biol Chem 1994; 269:16896-16901

89. Lortenkamp A. Are cadmium and other heavy metal compounds acting as endocrine disruptors? Metal Ions in Life Sci 2011; 8:305-317

90. Zhang RY, Liu Y, Pang DW, et al. Spectroscopic and voltametric study on the binding of aluminum (III) to DNA. Japan Soc Anal Chem, 2002; 18:761-766

91. Ionescu JG, Novotny J, Stejskal V, et al. Increased levels of transition metals in breast cancer tissue. Neuroendocrinol Lett 2006; 27(Suppl 1):36-39

92. Wu HDI, Chou SY, Chen DR, et al. Differentiation of serum levels of trace elements in normal and malignant breast patients. Biol Trace Element Res 2006; 113:9-18

93. Silva N, Peiris-John R, Wickremasinghe R. Cadmium a metalloestrogen: are we convinced? J Appl Toxicol 2012; 32:318–332 DOI 10.1002/jat.1771

94. Stoica A, Katzenellenbogen BS, Martin MB. Activation of estrogen receptor alpha by the heavy metal cadmium. Mol Endocrinol 2000; 14:545-553

95. Parodi DA, Greenfield M, Evans C, et al. Alteration of mammary gland development and gene expression by in utero exposure to arsenic. Reprod Toxicol 2015; In Press DOI 10.1016/j.reprotox.2014.12.011

96. Pollack AZ, Buck-Louis GM, Chen Z, et al. Trace elements and endometriosis: The ENDO study. Reprod Toxicol 2013; 42:41-48 DOI 10.1016/j.reprotox.2013.05.009

97. Darbre PD. Environmental Contaminants: Environmental Estrogens – Hazard Characterization. Encyclopedia of Food Safety 2014; 2:323–331 DOI 10.1016/B978-0-12-378612-8.00196-7

98. Kiyama R, Zhu Y, Kawaguchi K, et al. Estrogen-responsive genes for environmental studies. Environ Technol Innov 2014; 1-2:16-28 DOI 10.1016/j.eti.2014.09.001

99. Chighizola C, Meroni PL. The role of environmental estrogens and autoimmunity. Autoimmun Rev 2012; 11:A493-A501 DOI 10.106/j.autrev.2011.11.027

100. Wessels D, Barr DB, Mendola P. Use of biomarkers to indicate exposure of children to organophosphate pesticides: Implications for a longitudinal study of children's environmental health. Environ Health Persp 2003; 111(16): 1939-1946 DOI 10.1289/EHP.6179

101. Pimental D. Environmental and economic costs of application of pesticides primarily in the United States. Environ Dev Sustain 2005; 7:229-252. DOI 10.1007/s10668-005-7314-2

102. Food and Agriculture Organization of the United Nations (FAO), International Code of Conduct on the Distribution and Use of Pesticides. 2005; ISBN 92-5-105411-8 (accessed Feb 2, 2015)

3

Environmental Exposure and Health Effects Associated with Malathion Toxicity

1. Introduction

Organophosphate (OP) pesticides are a group of chemicals that have many domestic and industrial uses; historically, they have been most commonly used as insecticides and have been responsible for a number of pesticide poisonings. Accounting for about 70% of pesticide use in the United States, OPs have become the most commonly used pesticides because of the high persistence, accumulation, and toxicity of organochloride insecticides such as DDT and BHC. OPs are phosphorous-containing insecticides that were originally developed in the 1940s as highly toxic biological warfare agents. This group of chemicals includes insecticides such as malathion, diazinon, chlorpyrifos, methyl parathion, and parathion [1]. These compounds were first used in Germany during World War II as toxic nerve agents. Their modern derivatives include highly neurotoxic agents such as sarin, soman, and tabun. The main mechanism of toxic action of OPs is the inhibition of acetylcholinesterase enzyme activity, causing nervous and respiratory damages that may potentially result in death [2, 3]. It was not until World War II that the magnitude of detrimental effect on organisms was discovered from the research conducted to determine the toxicity of nerve gases used for military purpose [1]. Although they were produced during the World War II era, nerve gases were not used until the Iran–Iraq War and during an incident in Tokyo, Japan. During the Iran–Iraq War (1981–1988), it was reported that Iraq used nerve agents such as tabun and sarin. In March 1995, Aum Shinrikyo, a religious cult in Japan used bags of sarin on a subway train in Tokyo. The released gas killed 12 individuals and sent more than 5,000 to the hospital. They are generally lipid-soluble and are capable of penetrating the skin, the blood brain barrier, the placenta, and into the fetus [4].

2. Physical and Chemical Properties of Malathion

As an OP insecticide, malathion was first registered for use in the United States in 1956 by the United States Department of Agriculture (USDA). It is currently regulated by United States Environmental Protection Agency [1]. Malathion is a broad-spectrum insecticide used to control a variety of outdoor insects in both agricultural and residential settings. Estimation by the U.S. EPA indicates that over 30 million pounds of malathion are used annually [5]. About 60 percent is often used in federal and state programs to eradicate insects such as boll weevils, grasshoppers, and fruit flies. It is also used as a potent insecticide for mosquito control in residential areas as well as for insect control on a variety of food crops. Malathion has also been approved by the United States Food and Drug Administration (FDA) for addition in shampoos in order to control head lice [1]. Signal words ranging from "caution" to "danger" have been developed; depending on the combined toxicity of the active ingredient and other product components. Uses for individual malathion products vary widely; therefore, proper precautions should be taken to minimize their toxicity. The chemical structure of the technical grade of malathion is shown in Figure 1, and its physicochemical properties are presented in Table 1 [1, 3, 6, 7, 8].

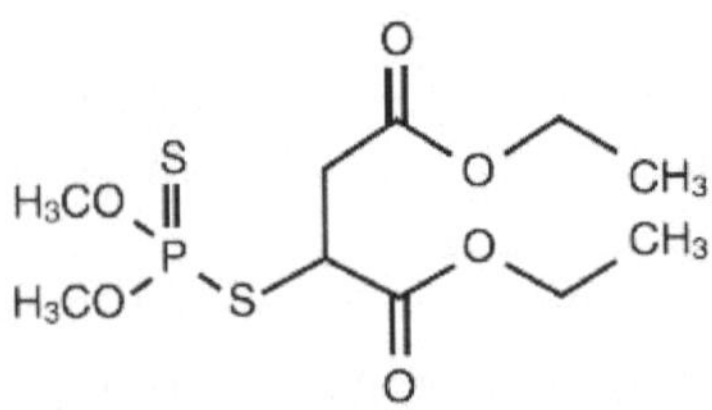

Figure 1. Structure of malathion

Table 1. Physical and chemical properties of malathion

Properties	Malathion	**Reference(s)**
CAS Reg. No.	121-75-5	[1]
Synonyms	Dimethyl dithiophosphate	[1]
Molecular Weight	of diethyl mercaptosuccinate	[6]
Color	330.4 g/mol	[3]
Physical State	Colorless to amber	[3]
Melting Point	Liquid	[3]
Boiling point	2.85°C	[7]
Density/Specific gravity	156°C	[7]
Odor	1.2076	[3]
Solubility Water (mg/L)	Skunk/garlic like	[6]
Vapor pressure (mmHg)	145 mg/L	[3, 6]
Henry's Law Constant	1.78×10^{-4} mmHg at 25°C	[8]
Soil Sorption Coefficient (K_{oc})	$2.0 (+1.2) \times 10^{-7}$	[3, 7]
	30, 93–1800	
	depending on soil type	
	and environmental conditions	

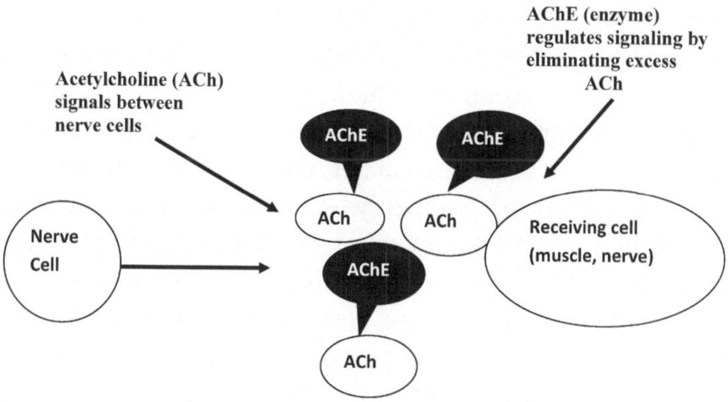

Figure 2. Normal nerve signals

3. Sources/Environmental Occurrence

OP introduced in the 1930s, are manufactured chemical substances that are produced by the reaction of alcohols and phosphoric acid. Their primary effect as insecticide was discovered during military operations when initially used as nerve gases [9]. Malathion, an OP compound, is also known as carbophos, maldison, and mercaptothion. Being a non-systemic, wide-spectrum insecticide, malathion is one of the most frequently used OP pesticides. It has been used for various eradication programs and for public health purposes throughout the United States and other countries. Some of the common areas of usage include agricultural, industrial, and use by the general public.

Malathion was developed during World War II, in the 1950s, and has been known for its high insect potency, but low mammalian toxicity. Considered to be one of the safest OP compounds, malathion is known as one of the most selective OP insecticides. It has been effective in the control of pests on vegetables, field crops, fruits, agriculture, commercial extermination, fumigation, domestic animals, and veterinary practices. Hence, it has many applications in agricultural, nonagricultural, and public health purposes. It is commonly used on agricultural crops (alfalfa, apple, apricot, asparagus, avocado, barley, bean, beats, blackberry, blueberry, boysenberry, broccoli, cabbage, carrot, cauliflower, celery, chayote, cherry, clover, corn, cotton, cucumber, dewberry, eggplant, potato, fig, garlic, grape, grapefruit, hay grass, horseradish, leek, lemon, lettuce, lime, loganberry, mango, pepper, pineapple, pumpkin, reddish, raspberry, spinach, wheat, squash, strawberry, tangerine, tomato, walnut, watermelon, wild rice, yam, and indoor-stored commodity treatment and empty storage facilities for barley, corn, oats, and wheat), stored products, golf courses, home gardens, trees and shrubs, mosquito control, Mediterranean fruit flies (medflies), fleas on pets, treatment of head lice (humans), household insects, Boll Weevil Eradication program, Christmas trees, lawn, etc. [1,3,4]. Homeowners use malathion for the following purposes: on ornamental flowers, shrubs, and trees, outdoor garbage dumps, irrigation and sewage systems, pastures, and range land. It can also be used to control ectoparasites of cattle, flies, and human head and body lice [3]. On the other hand, malathion's targeted pests include: ants, aphids, apple mealybug, armyworm, bagworm, beetle, borer, bug, fireworm, blueberry maggot, caterpillars, cattle lice, cockroaches, cherry fruitworm, rootworms, cotton fleahopper, cotton leafworm, cranberry fruitworm, European fruit lecanium, fleahoppers, fleas, flies, grasshoppers, green cloverworm, imported cabbageworm, leafhoppers, mosquitoes (adult, larvae), moths, mushroom flies, orangeworms, pepper maggot, pickleworm, plant bugs, poultry lice, sawflies, scales, spiders, ticks, tomato fruitworm, wasps, weevil, etc. [1,3,4]. Malathion is regulated by both FDA and U.S. EPA at a maximum amount of 8 parts per million (ppm) as residue on specific crops used for food. Because of its potential toxicity to humans, the EPA requires that an appropriate time lapse be observed between the application time and entry/reentry of a field worker.

4. Uses and Environmental Exposure

Malathion is an OP insecticide that is used mostly in agriculture and in public health programs to control infestations of insects including ants, aphids, fleas, fruit flies, hornets, mites, mosquitoes, moths, spiders, thrips, ticks, wasps, and weevil. It is also used as pest control for agricultural food and feed crops including blueberries, raspberries, strawberries, limes, cotton, cherries, garlic, greens, dates, and celery [1]. In addition to the use of malathion in plant applications, it is a key component of personal hygiene products used for lice control [10]. Currently, malathion is still used in a large scale in agricultural sector and public health programs all over the world. The estimated average annual total domestic usage of malathion in the USA is approximately 15 million pounds of malathion as an active ingredient [11]. Between late 1970s and 2008, malathion was the primary pesticide used in the USDA Boll Weevil Eradication Program to protect cotton crops in the southern United States [12, 13]. In 1998, malathion and diazinon were applied in some areas of the state of Florida after an outbreak of Mediterranean fruit flies called Medflies. Medfly outbreak resulted in a significant reduction in agricultural yields. To minimize its damage, federal and state authorities implemented the Medfly Eradication Program. Within 5 months of application, 123 people reported symptoms associated with pesticide exposure, such as respiratory distress, gastrointestinal distress, neurological problems, skin reaction, and eye distress [14]. The United States used malathion among the insecticides to control mosquitoes carrying West Nile Virus during the year 2005 [15].

Published research has reported cases of malathion poisoning associated with accidental and/or intentional exposure to malathion. A previous study conducted in Japan reported 10 deaths out of 63 cases of accidental exposure to malathion, as well as 404 deaths out of 480 cases of malathion-associated suicides or homocides [16]. Other accidental death from malathion toxicity in human population has also been documented [17]. The median lethal dose (LD_{50}) of malathion is estimated to be 2100 mg/kg in man [18].

Exposure to malathion occurs via dermal contact, ingestion, and/or inhalation [6]. Most people are not exposed to malathion in the air that they breathe or on things that they touch, unless they live near areas being sprayed. The people who are at the greatest risk to malathion toxicity are those who are occupationally exposed. These include farm workers, chemical sprayers, and people who work in factories that make malathion or other malathion-containing products. These high-risk groups can be exposed through skin absorption by contacting contaminated products or surfaces, or through lung absorption by inhaling contaminated air. Domestic users of malathion are also at high risk of intoxication related to its application in residential areas near homes and gardens for medflies and mosquitoes control. Exposure to high concentrations of malathion has been associated with severe toxicity and death in some cases. Hence, it has been recommended not to enter or go to the fields sooner than 6 days after malathion spraying [4]. Also, the utilization of personal protective devices such as breathing equipment and special clothings may prevent toxicity and protect against malathion intoxication. A previous study conducted by the U.S. EPA between 1971 and 1991 in 3 states pointed out that malathion was the only chemical detected in twelve groundwater monitoring wells. The highest malathion concentration of 6.17 ppb was reported in Virginia in a county where the land was mainly agricultural and forested.

More recent investigations on the environmental contamination have reported the absence of malathion in groundwater near areas that have been chemically sprayed; indicating a lower risk of malathion toxicity in drinking water collected from groundwater. Symptoms of exposure to malathion include headache, nausea and vomiting, burning eyes, difficulty breathing, and lethargy. Malathion exposure has been associated with metabolic disorders [19], oxidative stress [20], immunotoxicity [21], inflammation [22], and hepatotoxicity [23]. Malathion has also been reported to induce genetic damage in a variety of laboratory studies, including a study of mice fed with malathion-treated grains. In epidemiological studies with human blood cells, DNA damage and oxidative stress have been proposed as a process that could mechanistically link pesticide exposure to a number of health outcomes [2]. According to the U.S. EPA, there is evidence that malathion causes cancer. Experimental studies have pointed out that the commercial grade of malathion insecticide causes breast cancer in laboratory animals. Also, the use of malathion by farmers has been associated with an increased incidence of non-Hodgkin's lymphoma [24].

5. Environmental Fate and Transport

In general, OPs are transported through the environment in various ways. Malathion released in the atmosphere as a result of its use on agricultural crops and/or residential areas may form droplets that fall on ground covers including plants, animals, soils, water resources, buildings, and/or other structures. Malathion deposited on these platforms may subsequently be transported away through the action of rainfall/precipitation, and wind. It has been reported that malathion may remain in the environment for up to few months, but is usually transformed or degraded within a few weeks through the processes of photolysis, hydrolysis, and/or biodegradation by microorganisms. Because malathion is rapidly degraded by soil bacteria, low concentrations are expected to be present in groundwater [4]. Reported half-lives in soil range from 1 to 17 days [25, 26]. In water, malathion breaks down quickly by hydrolysis or by the action of bacteria present in the water. The half-lives of malathion in water were estimated as 1.65 days at pH 8.16, and 17.4 days at pH 6.0 [27]. In air, malathion is broken down by reacting with other chemicals formed naturally in the air by sunlight, to form a more toxic product called malaoxon. A study conducted in the Sierra Nevada Mountains reported very low malathion concentrations in air (< 1 ng/m^3), and concentrations ranging from 64 to 83 ng/L in surface waters between 18 and 2042 altitude. These results led the investigators to speculate that the distribution of malathion was a result of atmospheric transport [28]. If malathion is present on dry soil or on man-made surfaces such as sidewalks, pavements, or playground equipment, it usually does not break down as fast as it would in moist soil.

Published data indicate that malathion may be transported in the air following application to either agricultural or urban/residential areas [28, 29]. Malathion may be transported in the atmosphere as a vapor or adsorbed onto particulate matter [30]. Also, its occurrence in the atmosphere is generally localized. However, in a non-U.S. study of malathion adsorbed to fly ash (particulate matter) [31]. Adsorbed malathion is photodegraded when exposed to irradiation for up to 1.5 hours, but does not degrade when adsorbed to kaolin. The results from this study indicated that malathion adsorbed to kaolin maybe transported over long distances, while that adsorbed to fly ash will be photodegraded and therefore will not be transported far in the atmosphere [30]. Additionally, malathion has been detected in the fog of remote pristine areas, indicating that long-range transport may occur under some conditions [32].

6. Toxicokinetics

6.1. Absorption and distribution

Absorption of malathion occurs through the gastrointestinal tract, the respiratory tract and its primary and slowest absorption pathway, and the skin. Ingestion of contaminated food or water is the predominant route of exposure to malathion for the general population, compared to the inhalation and dermal routes. The predominant route of occupational exposure for the general population is through the dermal contact. Although it is well known that malathion is rapidly absorbed through the gastrointestinal tract and the skin, little is known about its fate from inhalation exposure. Absorbed malathion can be transported by the blood and distributed to many organs and tissues including the liver where it is metabolized to form malaoxon. In biologic systems, malathion and its metabolites have a very low accumulation potential and are eliminated through urine within a few days. Hence, analysis of malathion or its metabolites in urine should be performed within few days after exposure. Their concentrations in tissues and body fluids are important biomarkers of exposure. Currently, there is a scarcity of scientific data regarding the background concentrations of malathion in human tissues [4].

Dermal absorption of malathion is rapid. However, the absorption rate vastly depends on it applied dose and the exposure site [4]. From a study examining the absorption rates of applied malathion on various parts (forearm, axilla (armpit), ball of the foot, abdomen, forehead, and jaw angle) of the skin of male human volunteers, the greatest rate of absorption was found in the armpit followed by the forehead. The armpit and forehead areas respectively showed 4.2 and 3.4 times greater absorption than the forearm skin [33]. In another toxicokinetics study, it was reported that more than 90% of absorbed malathion was eliminated through urine within 24 hours of exposure, by male rats orally exposed to 28 mg/kg malathion, or dermally exposed to 41 mg/kg malathion. The remaining malathion was detected in the feces, blood, intestines, liver, and kidneys [34]. Based on organ weight changes during a two-week inhalation study in rats, other target organs for malathion distribution and toxicity included the liver and the kidney [35].

6.2. Mode of action

OP pesticides including malathion share a common mode of action. They bind to the enzyme acetylcholinesterase (AChE) at nerve endings throughout the bodies of insects and other organisms [36]. AChE plays a key role in the synaptic transmission of nerve impulses. Its inhibition causes the blockage to signal transmission leading to intoxication manifested by restlessness, hyperexcitability, convulsions, paralysis, and death [24]. A similar mode of action has been reported for all OP insecticides [37]. Malaoxon, the toxic metabolite of malathion, is known to illicit a similar effect in mammals. However, the signs and symptoms of malathion toxicity are different in mammals and insects because in mammals AChE is not active in the central nervous system, but rather in nerves that connect with muscles [37]. Malathion is toxic via skin contact, ingestion, and inhalation exposure [6]. Under normal circumstances, AChE binds to the neurotransmitter acetylcholine (ACh) at the nerve junction, effectively ending the stimulation of the next neuron. Resulting effects from malathion toxicity include restlessness, hyperexcitability, convulsions, blurred vision, salivation, difficulty breathing, chest tightness, diarrhea, vomiting, sweating, headaches, and cramps [4, 24]. Intermediate syndrome (delayed neuropathy) has been reported in humans as a result of acute exposure to high amounts of malathion. Symptoms include weakness in several motor cranial nerves, weakness in neck flexors and proximal limb muscles, and respiratory paralysis.

There is evidence that exposure to malathion below the level that causes nervous system effects results in few or no health problems. Toxicity is usually the result of binding of AChE to malaoxon (malathion metabolite) which leads to the accumulation of ACh at the nerve junctions and subsequent overstimulation of the nervous system [36]. Malaoxon, the primary toxic metabolite of malathion, is produced in the liver as a result of a biotransformation process involving an oxidative sulfuration catalyzed by the cytochrome P450 enzyme [38, 39]. Findings from experimental studies have pointed that malaoxon is 22 times more toxic than malathion when exposure is by oral route, and 33 times more potent by all routes of exposure from acute and sub-acute exposure durations [40]. Exposure to multiple OPs can lead to additive toxicity. However, the different OPs vary widely in their potency and how well they are absorbed by the body depending on the route of exposure [36].

The accumulation of acetylcholine at the nerve junctions as a result of OP binding to the acetylcholinesterase enzyme, phosphorylating its serine hydroxyl group, and deactivating its functional esterase site [9]. The buildup of acetylcholine at the neuromuscular junction also causes a persistent depolarization of the skeletal muscle, leading to fasciculations, tremors, ataxia, weakness, convulsions, and coma. [41]. In the central nervous system, neural transmission is disrupted. Acetylcholinesterase is also contained in the erythrocytes, and is identical to that which is found in the nervous system; however, the function is to control, to a certain extent, permeability of the cell membrane.

Malathion toxicity causes a disruption of the functioning of the cholinergic system, and elicits basic clinical signs and symptoms that are similar in humans and other mammals [4]. Both muscarinic and nicotinic receptors as well as central nervous system receptors are modulated by malathion exposure [36]. Several muscarinic effects such as salivation, lacrimation (production of tears), urination and defecation (the SLUD syndrome), vomiting, dyspnea (shortness of breath), bradycardia (reduced heart rate), abdominal pain, miosis (constriction of the pupils), and anorexia have been documented as a result of malathion interaction with AchE and over excitation of the post-ganglionic parasympathetic receptors in the nervous system [42]. Other clinical signs of toxicity including muscle tremors and rigidity, weakness and loss of limb mobility, and paralysis have been observed as a result of excessive stimulation of nicotinic receptors [40].

In humans, the clinical manifestations of malathion toxicity depend on several factors including the target enzyme and its sensitivity, the site of interaction at the synaptic junction, the dose of malaoxon that interacts with the receptor, and the exposure route [4]. Several muscarinic effects including excessive perspiration, constriction of the pupils, lacrimation, salivation, abdominal cramps, diarrhea, nausea, vomiting, chest tightness and difficulty breathing have been reported in humans [4, 36, 43].

6.3. Metabolism and distribution

From an in vivo study examining the metabolism and distribution of malathion, ten metabolites were found in the urine and feces of rats pre-exposed to radiolabeled malathion. A large amount of radiolabeled compound (80%) was excreted in urine, and was comprised mainly of malathion dicarboxylic acid and of thiomalic acid and malathion mono acids to a lesser extent. Malaoxon, desmethyl malathion, O,O-dimethyl phosphorothioate, monoethyl fumarate, and thiomalic acid concentrations were relatively low [38]. Similar metabolites of malathion were reported in both humans and rats, with the exception of monomethyl and dimethyl phosphate that were detected in humans, and of thiomalic acid and monoethyl fumarate that were found in rats. An experimental study with rats also reported that malathion had low accumulation affinity in tissues, and constituted the majority of residual compounds excreted [44].

It has been pointed that other components in malathion-containing products can enhance its toxic action by deactivating the activity of the carboxyesterase enzymes that catalyze its conversion to malaoxon [45]. These other constituents of malathion compounds are impurities that may result from contamination during manufacturing and/or chemical storage [4]. Through the action of carboxylesterases, both malathion and malaoxon are degraded into metabolic products that are more water-soluble and less toxic [44]. In humans, carboxylesterases are not found in the blood, but are mainly present in the liver which constitutes the main organ of biotransformation. In rats, they have been found in many organs including the liver, blood serum, and kidney [4]. Secondary metabolic pathways include oxidative desulfuration to malaoxon, hydrolysis to phosphatases, and dealkylation to desmethylmalathion [38].

6.4. Excretion

Orally administered malathion has been reported to be excreted in large amount (80–90% as parent compound) in the urine within the first 24 hours post-exposure[44]. From a toxicokinetics study with radiolabeled malathion, it was reported that ten different metabolites of malathion were detected in the urine of rats. Urinary excretion accounted for about 85–89% of the exposed dose, while fecal excretion accounted for about 4–15% within 72 hours post-exposure[38]. In a study examining the dermal absorption, metabolism, and excretion, malathion was applied to the ventral forearm skin of eight human male volunteers at 4 $\mu g/cm^2$. Its excretion was highest at 4–8 hours post-exposure; however, only 8% of its initial dose was found in urine within 120 hour post-exposure [46]. Malathion has been detected in human breast milk [47], although no studies were found that examined the relationship to exposure or if its presence could cause adverse effects in nursing infants.

7. Genotoxicity Studies

Malathion has been identified by the National Institute for Occupational safety and Health (NIOSH) as a mutagen, based on a comprehensive review of scientific evidence from many mutagenicity tests including bacteria, fruit flies, mice, hamsters, fish and human cell cultures bioassays conducted in 29 laboratories between 1978 and 1995 [48]. Similar findings have been reported from recent studies, supporting those that were previously reviewed by NIOSH. From an in vivo genotoxicity study, researchers from Assan and North-Eastern Hill Universities (India) demonstrated that oral exposure to malathion induces genetic damage in mice [49]. Other investigators from the Egyptian National Research Center reported that mice fed with malathion-treated wheat developed genetic damage of two different types at all tested doses [50]. Many other investigations have shown that exposure to malathion, its metabolite malaoxon, and its contaminant isomalathion induces genetic damage in human blood lymphocytes [51–53].

8. Carcinogenicity Studies

Evidence of the carcinogenic effects of few pesticides in animals and an increase in the risk of developing malignancies in occupationally exposed populations have made necessary studies in exposed workers [54–58]. Several studies have been conducted with rats and mice to determine whether malathion has the potential to cause cancer, with variable results. In April 2000, U.S. EPA classified malathion as having "suggestive evidence of carcinogenicity but not sufficient to assess human carcinogenic potential by all routes of exposure". This categorization was made considering Cheminova's findings that malathion induces hepatocellular carcinomas in experimental animals at higher doses of exposure. [59].

Recent studies have underscored the need to strengthen EPA's carcinogenicity classification. From a 2001 study conducted by researchers at Columbia University and the Universities of Tarapaca and Concepcion (Chile), malathion exposure significantly increases the incidence rate of breast cancer in rats [60]. Findings from another study demonstrated that the increase in cancer incidence was linked to the damage of an important gene by malathion [61]. In an investigation involving long-term dietary exposures to malathion, researchers observed an increased incidence of liver and nasal/oral tumors in rats and increased incidence of liver tumors in mice [44]. In a two-year dietary study, researchers administered oral doses of 2,359, 739, or 868 mg/kg/day to rats. They found no evidence of an association between tumor incidence and exposure to malathion [10, 62]. The International Agency for Research on Cancer (IARC) concluded in 1987 that the carcinogenic potential of malathion was not classifiable, and placed it in group 3 [63]. Also, an epidemiological study at six Canadian provinces pointed out that the cancer risk of non-Hodgkin's lymphoma was twice in men exposed to malathion compared to healthy men who had not been exposed to malathion [64]. This finding was consistent with the results of previous studies conducted in the United States. [65]. Also, occupational exposure to pesticides has been reported to be associated with an increase risk or incidence of different types of carcinomas such as non-Hodgkin's lymphoma [66], Hodgkin's lymphoma [66], leukemia [57], multiple lymphoma [67], pancreatic cancer [68], gastric cancer [69], lung cancer [70], bladder and colon cancer [71], and gall bladder sarcoma [72].

In addition to its potential genotoxic and carcinogenic effects, malathion has also been reported to have significant adverse effects on different organ systems. Its potential systemic health effects on specific organs include the following:

9. Hematologic Effects

At high doses, malathion acts like other OP insecticides to suppress the immune system in certain animal species [73]. A study indicated that malathion usage may affect the hematopoietic system [74]. In addition, the study reported that sublethal doses of malathion exposure caused deleterious effects on hematological parameters of treated animals [75]. Other studies indicated that chronic exposure to malathion significantly decreases RBCs, Hb, and P.C.V% values in treated animals compared to the control group [76]. Furthermore, similar studies demonstrated that high levels of malathion induced DNA abnormalities in exposed persons [77], decreased human immunity [78], and caused non-Hodgkin's lymphoma [79, 80]. Acute malathion treatment resulted in bone marrow failure and plastic anemia [81, 82]. Although acute toxicity study with malathion displays deleterious effects in humans and test animals, chronic toxicity study revealed that a group of volunteers who ingested low dose of malathion over a period of 1.5 months did not show a significant inhibition of their blood cholinesterase activity [16].

10. Neurologic Effects

Malathion is a widely used OP insecticide because of its relatively low toxicity to mammals and its high selectivity compared to other OP insecticides. Possible symptoms of acute exposure to high levels of malathion include skin and eye irritation, cramps, nausea, diarrhea, excessive sweating, seizures and even death. Most symptoms tend to resolve within several weeks. According to EPA, there is currently no reliable information on adverse health effects of chronic exposure to malathion [83]. Malathion is usually less toxic. However, its overall toxicity is influenced by its metabolites and other chemical constituents of its chemical formulation. Malathion oxidative metabolism results in the production on malaoxon in mammals, insects, and plants. Being the most hazardous form, malaoxon is 40 times more acutely toxic than malathion [84, 85]. Interestingly, malathion present in the body system will clear up or be eliminated within three to five days [86]. The rapid rate of excretion from human body is facilitated by the action of carboxyesterases that catalyze the biotransformation of malathion and its metabolites to non-toxic and water-soluble products that can be easily eliminated from the body or cells. Arthropods such as insects lack or possess a low level of carboxyesterases. Therefore, insects are highly sensitive to malathion toxicity. Today, malathion is still considered as one of the safest OP insecticides. It was used for large eradication programs against insect infestations in metropolitan areas of Florida, Texas, and California [87].

11. Reproductive and Developmental Effects

Malathion is known to influence the reproductive function through two mechanisms including its cellular toxic action and its effect on the encephalic regulatory serotoninergic, besides acetylcholinergic which is the main mechanism of the reproductive functions [88]. The enzyme acetylcholinesterase (AChE) is used as a marker for exposure to OPs and carbamates (both inhibit this enzyme, resulting in a general nervous system failure) [89]. Malathion is found to inhibit the release of acetylcholinesterase at the synaptic junction [60]. AChE plays an important role in the control of nerve excitability at post-synaptic sites. Inhibition of liver AChE activity is a useful indicator of OP pesticides poisoning. In addition, many scientific reports indicated that malathion-induced physiological, biochemical, immunological, and histological changes in experimental animals [90–92]. A documented scientific report showed that high doses of malathion induce developmental and reproductive effects in experimental animals [93]. Another report indicated malathion and its metabolites can cross the placenta of mammals and depress cholinesterase activity of the fetus [94]. A 2003 report indicated that malathion reduced sperm count and the number of normal forms in test mice with a maximal effect at 18 days post-injection [95]. Previous studies indicated that administration of malathion caused damage to the Leydig cells and decreased the levels of testosterone [96, 97]. Similar study showed reduction in the number of immature germ cells due to decrease of steroidogenic activity and damage of the Sertoli cells [96]. Another study demonstrated that malathion interferes with the process of spermatogenesis by preventing the maturation in the later post-meiotic stages, which are androgen-dependent [98].

12. Hepatic Effects

There is limited data in the literature regarding the hepatic effects of malathion in agricultural workers. However, recent scientific data have demonstrated that malathion and other pesticides induce liver and kidney histopathological alterations in experimental animals [92, 99–101]. The study conducted showed that malathion intoxication may affect the structures of the liver and kidney showing the presence of fine subcapsular infiltrations, diffused parenchymatous degeneration of single hepatocytes, presence of fine foci constructed of plasmatic cells, and histiocytes located between hepatic plates [102].

13. Regulatory Guidelines

Malathion is an OP insecticide that was first registered in 1956, and remains largely in use worldwide. The regulations and recommendations for malathion include the following: The Food and Drug Administration (FDA) and the Environmental Protection Agency (EPA) allow a maximum amount of 8 parts per million (ppm) of malathion to be present as a residue on specific crops used as foods [10]. The Occupational Safety and Health Administration (OSHA) has established an exposure limit for malathion in the workplace of 15 milligrams per cubic meter (mg/m^3), for an 8-hour workday, 40 hours per week [79]. According to the National Institute for Occupational Safety and Health (NIOSH)'s guidelines, workers should not be exposed to malathion concentrations greater than 10 mg/m^3 during a -10 hour workday, 40 hours per week. NIOSH also recommends that an atmospheric concentration of 250 mg/m^3 malathion be considered as being immediately hazardous to human health and life.

References

1. United States Environmental Protection Agency (U.S. EPA): Reregistration Eligibility Decision (RED) for malathion; EPA 738-R-06-030; U.S. Environmental Protection Agency, Office of Prevention, Pesticides and Toxic Substances, Office of Pesticide Programs, U.S. Government Printing Office: Washington, DC. 2006.
2. Ojha A, Srivastava N. In vitro studies of organophosphate pesticides induced oxidative DNA damage in rat lymphocytes. Mutat Res 2014;761:10–17.
3. Hazardous Substances Databank (HSDB): Malathion; U.S. Department of health and Human Services, National Institutes of Health, National Library of Medicine. Updated June 2005. http://toxnet.nlm.nih.gov/cgi-bin/sis/htmlgen?HSDB [accessed: Jan 2008].
4. Agency for Toxic Substances and Disease Registry (ATSDR): Toxicological Profile for Malathion; U.S. Department of Health and Human Services, Agency for Toxic substances and disease registry: Atlanta, 2008.
5. Donaldson DT, Kiely AG. Pesticides industry sales and usage: 1998 and 1999 market estimates. U.S.EPA. Office of Pesticide Programs. www.epa.gov/oppbead1/pestsales. 2002, pp.14–5.
6. Tomlin CDS. The Pesticide Manual, a World Compendium. 14th edn. British Crop Protection Council. Alton, Hampshire, UK. 2006; 642–3.
7. Hornsby AG, Wauchope RD, Herner AE. Pesticide Properties in the Environment; Springer-Verlag: New York, 1996.
8. Fendinger NJ, Glotfelty DE. Henry's law constants for constants for selected pesticides, PAHs and PCBs. Environ Toxicol Chem 1990;9:731–5.
9. Dyro FM. Organophosphate. eMedicine. Updated 13 March 2003. Available from: http://www.emedicine.com/neuro/topic286.htm [accessed: 7 July 2005].

10. Agency for Toxic Substances and Disease Registry (ATSDR): Toxicological profile for malathion. U.S. Department of Health and Human Services, Agency for Toxic substances and disease registry: Atlanta, GA, 2003.
11. United States Environmental Protection Agency (U.S. EPA): Registration and eligibility decision (RED) for malathion; United States Environmental Protection Agency (EPA 738-R-06-030). 2008.
12. United States Department of Agriculture Animal and Plant Health Inspection Service: Questions and answers. The EPA's risk assessment on malathion. 2006.
13. United States Department of Agriculture Animal and Plant Health Inspection Service: Cotton pests. 2008.
14. Center for Disease Control (CDC): Surveillance for acute pesticide-related illness during the Medfly Eradication Program-Florida 1998. 1999.
15. Extension Toxicology Network: Pesticide Information Profile-Malathion. 1996.
16. Agency for Toxic Substances and Disease Registry (ATSDR): Toxicologic information about insecticides used for eradicating mosquitoes (West nile virus control). 2005.
17. Matsumura F. Hazards to man and domestic animals, in: Toxicology of Insecticides, Plenum Press, New York. 1975;411–2.
18. Uygun U, Koksel H, Atli A. Residue levels of malathion and its metabolites and fenitrothion in post-harvest treated wheat during storage, milling and baking. Food Chemistry 2005;92:643–7.
19. Lasram MM, Annabi AB, Elj N, Selmi S, Karmoun A, El-Fazaa S, Gharbi N. Metabolic disorders of acute exposure to malathion in adult Wister rats. J Hazard Mater 2009;163:1052–5.
20. Alp H, Aytekin I, Hatipoglu NK, Alp A, Ogun M. Effects of sulforophane and curcumin on oxidative stress created by acute malathion toxicity in rats. Eur Rev Med Pharmacol Sci 2012;16(3):144–8.
21. Nain S, Bour A, Chalmers C, Smits JE. Immunotoxicity and disease resistance in Japanese quail (Coturnix coturnix japonica) exposed to malathion. Ecotoxicology. 2011;20:892–900.
22. Mostafalou S, Eghbal MA, Nili-Ahmadabadi A, Baeeri M, Abdollahi M. Biochemical evidence on the potential role of organophosphates in hepatic glucose metabolism toward insulin resistance through inflammatory signaling and free radical pathways. Toxicol Ind Health 2012;28(9):840–51.
23. Kalender S, Uzun FG, Durak D, Demir F, Kalender Y. Malathion induced hepatotoxicity in rats: the effects of vitamins C and E. Food Chem Toxicol 2010;48:633–8.
24. Winter. Insecticide Factsheet Malathion. J Pesticide Reform 2003;23(4).
25. Bradman A, Harnley ME, Goldman LR, Marty MA, Dawson SV, Dibartolomeis MJ. Malathion and malaoxon environmental levels used for exposure assessment

and risk characterization of aerial applications to residential areas of southern California. 1989–1990. J Expo Anal Environ Epidemiol 1994;1:49–63.

26. Getenga ZM, Jondiko JIO, Wandiga SO, Beck E. Dissipation behavior of malathion and dimethoate residues from the soil and their uptake by the garden pea (Pisum sativum). Bull Environ Contam Toxicol 2000;64:359–67.
27. Wang T. Assimilation of malathion in the Indian River estuary, Florida. Bull Environ Contam Toxicol 1991;47:238–43.
28. LeNoir JS, McConnell LL, Fellers GM, Cahill TM, Seiber JN. Summertime transport of current-use pesticides from California's Central Valley to the Sierra Nevada mountain range, USA. Environ Toxicol Chem 1999;18(12):2715–22.
29. Majewski MS, Capel PD. Pesticides in the atmosphere: distribution, trends, and governing factors. Chelsea, MI: Ann Arbor Press. 1995;pp.8–79.
30. Bossan D, Wortham H, Masclet P. Atmospheric transport of pesticides absorbed on aerosols I. Photo degradation in simulated atmosphere. Chemosphere 1995;30(1):21–9.
31. Mulla MS, Mian LS, kawecki JA. Distribution, transport, and fate of the insecticides malathion and parathion in the environment. Residue Reviews; Gunther FA, Gunther JD, eds; Springer-Verlag: New York. 1981.
32. Rice C. Pesticides in fogwater. Pestic Outlook 1996;7(2):31–6.
33. Maibach HI, Feldman RJ, Milby TH, Serat WF. Regional variation in percutaneous penetration in man. Arch Environ Health 1971;23:208–11.
34. Zeid MMA, El-barouty G, Adbdel-Reheim E, Blancato J, Dary C, El-Sebae AH, Saleh M. Malathion's disposition in dermally and orally treated rats and its impact on the blood serum acetylcholine esterase and protein profile. J Environ Sci Health, Part B. 1993;28(4):413–30.
35. Malathion: Updated Revised Human health Risk assessment for the Reregistration Eligibility Decision Document (RED); EPA-HQ-OPP-2004-0348-0004; U.S. Environmental Protection Agency, Office of Prevention, Pesticides and Toxic Substances, Office of Pesticide Programs, U.S. Government Printing Office: Washington, DC, 2005.
36. Reigart JR, Roberts JR. Organophosphate insecticides. Recognition and Management of Pesticide Poisonings, 5th edn; U.S. Environmental Protection Agency, Office of Prevention, Pesticides and Toxic Substances, Office of Pesticide Programs, U.S. Government Printing Office: Washington, DC. 1999;pp.34–47.
37. Ware GW. The Pesticide Book. Fresno CA: Thomson Publications 2000;pp.178–183.
38. Roberts TR. Metabolic Pathways of Agrochemicals-Part 2: Insecticides and Fungicides; The Royal Society of Chemistry: Cambridge, UK. 1998;pp.360–367.
39. Costa LG. Toxic effects of pesticides. Casarett and Doull's Toxicology: The Basic Science of Poisons, 7th edn; Klaassen CD (ed.) McGraw Hill Medical: New York. 2008;pp.883–930.

40. USEPA: Revised Reregistration Eligibility Decision (RED) for Malathion; EPA 738-R-06-030; U.S. Environmental Protection Agency, Office of Prevention, Pesticides and Toxic Substances, Office of Pesticide Programs, U.S. Government Printing office: Washington DC. 2009.

41. Slapper D. Toxicity, organophosphate and carbamate. eMedicine. Updated 29 December 1999. Available from: http//members.aol.com/DonationDrive/Organphos Toxmedicine.html [accessed: 7 July 2005].

42. Blodgett DJ. Organophosphate and carbamate insecticides. Small Animal Toxicology, 2nd edn; Peterson ME, Talcott PA. (eds) Elsevier Saunders: Saint Louis. 2006;pp.941–953.

43. Wagner SL. Diagnosis and treatment of organophosphate and carbamate intoxication. Occup. Med.: State of the Art Reviews. 1997;12(2):239–49.

44. Edwards D. Reregistration Eligibility Decision for Malathion. US Environmental Protection Agency – Prevention, Pesticides and Toxic Substances. EPA 738-R-06-030 journal. 2006;9.

45. WHO: Environmental Health Criteria 63, Organophosphate Insecticides: A General Introduction; International Programme on Chemical Safety, World Health Organization: Geneva, Switzerland. 1986.

46. Feldman RJ, Maibach HI. Percutaneous penetration of some pesticides and herbicides in man. Toxicol Appl Pharmacol 1974;28:126–32.

47. Sanghi R, Pillai MKK, Jayalekshmi TR, Nair A. Organochlorine and organophosphorous pesticide residues in breast milk from Bhopal, Madhya Pradesh, India. Hum Exp Toxicol 2003;22(2):73–6.

48. National Institute for Occupational Safety and Health (NIOSH): 2002. Registry of Toxic effects of Chemical Substances: Succinic acid, mercapto-diethyl ester, S-ester with O, O-dimethylphosphorodithioate. Available from www.cdc.gov/niosh/rtecs/wm82c80.html.27.

49. Giri S, Prasad SB, Giri A, Sharma GD. Genotoxic effects of malathion: an organophosphorus insecticide, using three mammalian bioassays in vivo. Mutat Res 2002;514:223–31.

50. Amer SM, Fahmy MA, Aly FAE, Farghaly AA. Cytogenetic studies on the effect of feeding mice with stored wheat grains treated with malathion. Mutat Res 2002;513:110.

51. Pluth JM, O'Neill JP, Nicklas JA, Albertini RJ. Molecular bases of hprt mutations in malathion-treated human T-lymphocytes. Mutat Res 1998;397:137–48.

52. Blasiak J, Jaloszynski P, Trzeciak A, Szyfter K. In vitro studies on the genotoxicity of the organophosphorous insecticide malathion and its two analogues. Mut Res 1999;445:275–83.

53. Blasiak J, Stankowska D. Genotoxicity of malaoxon: induction of oxidized and methylated bases and protective effect of a-tocopherol. Pest Biochem Physiol 2001;71:88–96.

54. Lucas D, Ferrara R, Gonzales E, Albores A, Manno M, Berthou F. Cytochrome CYP2E1 phenotyping and genotyping in the evaluation of health risks from exposure to polluted environments. Toxicol Lett 2001;124:71–81.

55. Mills PK, Zahm SH. Organophosphate pesticide residues in urine of farm workers and their children in Fresno County, California. Am J Ind Med 2001;40:571–7.

56. Catano HC, Carranza E, Huamani C, Hernandez AF. Plasma cholinesterase levels and health symptoms in Peruvian farm workers exposed to organophosphate pesticides. Arch Environ Contam Toxicol 2008;55(1):153–9.

57. Bonner MR, Williams BA, Rusiecki JA, Blair A, Beane Freeman LE, Hoppins JA, Dosemeci M, Lubin J, Sandler DP, Alavanja MC. Occupational exposure to terbufos and the incidence of cancer in the Agricultural Health Study. Cancer Causes Control 2010;21(6):871–7.

58. Mackenzie-Ross SJ, Brewin CR, Curran HV, Furlong CE, Abraham-Smith KM, Harrison V. Neuropsychological and psychiatric functioning in sheep farmers exposed to low levels of organophosphate pesticides. Neurotoxicol Teratol 2010;32(4):452–9.

59. USEPA: Office of Prevention, Pesticides and Toxic Substances. Malathion. Human Health risk assessment for the reregistration eligibility decision. Chemical no. 057701. Case No. 0248. Barcode D269070. Available from www.epa.gov/oppsrrd1/op/malathion.htm. 2000; p.1.

60. Cabello G, Valenzuela M, Vilava A. A rat mammary tumor model induced by the organophosphorous pesticides parathion and malathion, possibly through acetylcholinesterases inhibition. Environ Health Persp 2001;109:471–9.

61. Cabello G, Juarranz A, Botella LM, Calaf GM. Organophosphorous pesticides in breast cancer progression. J Submicrosc Cytol Pathol 2003;35:1–9.

62. Daly IA. 24-month oral toxicity/oncogenicity study of malathion in the rat via dietary administration. Final report: Lab project No.90-3641. 1996. Unpublished study prepared by Huntington Life Sciences. EPA MRID 4394201. Toxicological Profile for Malathion; U.S. Department of Health and Human Services, Agency for Toxic Substances and Disease registry, Public Health Service: Atlanta. 2003.

63. International Agency for Research on Cancer (IARC): Miscellaneous Pesticides. IARC Monographs on the Evaluation of Carcinogenic Risks to Humans; International Agency for Research on Cancer: Lyon, France. 1998;30:103.

64. McDuffie HH, Pahwa P, McLaughlin JR, Spinelli JJ, Fincham S, Dosman JA, Robson D, Skinnider LF, Chio NW. Non-Hodgkin's lymphoma and specific pesticide exposures in men: cross-Canada study of pesticides and health. Cancer Epidemiol. Biomarkers Prev 2001;10:1155–63.

65. Cantor KP, Blair A, Brown LM, Burmeister LF, Everett G. Pesticides and other agricultural risk factors for non-Hodgkin's lymphoma among men in Iowa and Minnesota. Cancer Res 1992;52:2447–55.

66. Orsi L, Delabra L, Monnereau A, Delval P, Berthou C, Fenaux P, Marit G, Sobeyran P, Huguet F, Milpied N, Leporrier M, Hemon D, Troussard X, Clavel J. Occupational

exposure to pesticides and lymphoid neoplasms among men: results of a French case-control study. Occup Environ Med 2009;66(5):291–8.

67. Baris D, Silverman DT, Brown LM, Swanson GM, Hayes RB, Schwartz AG, Liff JM, Schoenberg JB, Pottern LM, Greenberg RS, Stewart PA. Occupation, pesticide exposure and risk of multiple myeloma. Scand J Work Environ Health 2004;30(3):215–22.
68. Andreotti G, Freeman LE, Hou L, Coble J, Rusiecki J, Hoppin JA, Silverman DT, Alavanja MC. Agricultural pesticide use and pancreatic cancer risk in the Agricultural Health Study Cohort. Int J Cancer 2009;124(10):2495–500.
69. Mills PK, Yang RC. Agricultural exposures and gastric cancer risk in Hispanic farm workers in California. Environ Res 2007;104(2):282–9.
70. Beane Freeman LE, Bonner MR, Blair A, Hoppin JA, Sandler DP, Lubin JH, Dosemeci M, Lynch CF, Knott C, Alavanja MC. Cancer incidence among male pesticide applicators in the Agricultural Health Study Cohort exposed to diazinon. Am J Epidemiol 2005;162(11):1070–9.
71. Koutros S, Lynch CF, Ma X, Lee WJ, Hoppin JA, Christensen CH, Andreotti G, Freeman LB, Rusiecki JA, Hou L, Sandler DP, Alavanja MC. Heterocyclic aromatic amine pesticide use and human cancer risk: results from the U.S. Agricultural Health Study. Int J Cancer 2009;124(5):1206–12.
72. Shukla VK, Rastogi AN, Adukia TK, Raizada RB, Reddy DC, Singh S. Organochlorine pesticides in carcinoma of the gall-bladder: a case-control study. Eur J Cancer Prev 2001;10:153–6.
73. Dean JH, Murray MJ. Toxic responses of the immune system. In: Mary OA, Doull J, Klaassen C. (eds.) Casarett and Doull's Toxicology, the Basic Science of Poisons, Fourth Edition. Pergamon Press, NY. 1991.
74. Schalm OW, Jain NC, Carrol EJ. Veterinary Hematology Leo and Febiger, 3/E. 1975.
75. Jalel HA. The effect of malathion on the some hematological parameters of albino mice. Bas J Vet Res 2012;11(1): 246-253
76. ELZawahry EI. Assessment of toxicity on chronic treatment with some pesticides on albino rat. Bull Egypt Soc Physiol Sci 2004;24(1:251- 264.
77. Agency for Toxic Substances and Disease Registry (ATSDR): Toxicological profile for malathion draft for public comment Atlanta: US Department of Health and Human Services. 2006.
78. Banarjee BD, Keener DC, Ray A. Immunotoxicity of pesticides perspectives and trends. Indian J Exp Biol 1996;723–.
79. Agency for Toxic Substances and Disease Registry (ATSDR): Toxicological profile from malathion draft for public comment Atlanta: US Department of health and human Services. 2001.
80. Cantor KP, Blair A, Everett G, Gibson R, Burmeister LF, Brown LM, Schuman L, Dick FR. Pesticides and other agricultural risk factors for non-Hodgkin's lymphoma among men in Iowa and Minnesota. Cancer Res 1992;52(9):2447–55.

81. Gyton C, Hall E. Text Book of Medical Physiology, 11th edn. Elsevier Inc. Pennsylvania. US App. 2006;pp.272–276.
82. Zahm SH, Weisenburger DD, Saal RC, Vaught JB, Babbitt PA, Blair A. The role of agricultural pesticide use in the development of non Hodgkin's lymphoma in woman arch. Environ Health 1993;48(5):353–8.
83. US Department of Health and Human Services: Agency for Toxic Substances and Disease Registry. Medical Management Guidelines for Malathion. Retrieved 2008-04-02. 2008.
84. Brodeur J, DuBois KP. Studies on factors influencing the acute toxicity of malathion and malaoxon in rats. Canad J Physiol Pharmacol 1967;45(4):621–31.
85. Aldridge WN, Miles JW, Mount DL, Verschoyle RD. The toxicological properties of impurities in malathion. Arch Toxicol 1976;42(2):95–106.
86. Maugh II, Thomas H. Study links pesticide to ADHD in children. Los Angeles Times. 2010.
87. Flessel P, Quintana PJE, Hooper K. Genetic toxicity of malathion. A review. Environ Mol Mutagenesis 1990;22:7.
88. Uluitu M, Boca A, Petec G, Chis R, Catrinescu G. The influence of malathion on the brain serotonin and reproductive function in rats. Physiologie 1981;18:167–74.
89. Fulton MH, Key PB. Acetylcholinesterase inhibition in estuarine fish and invertebrates as an indicator of organophosphorus insecticide exposure and effects. Environ Toxicol Chem 2001;20:37–45.
90. Rezg R, Mornagui B, El-Arbi M, Kamoun A, El-Fazaa S, Gharbi N. Effect of subchronic exposure to malathion on glycogen phosphorylase and hexokinase activities in rat liver using native PAGE. Toxicology. 2006;223(1–2):9–14.
91. Rezg R, Mornagui B, Kamoun A, El-Arbi M, El-Fazaa S, Gharbi N. Effect of subchronic exposure to malathion on metabolic parameters in the rat. Comptes Rendus Biologies 2007;330(2):143–7.
92. Saadi L, Lebaili N, Benyoussi M. Exploration of cytotoxic effect of malathion on some rat organs structure. Commun Agri Appl Biol Sci 2008;73(4):875–81.
93. Gallo MA, Nicholas JL. Organic phosphorous pesticides. In: Wayland JH, Edward RL (eds.) Handbook of Pesticide Toxicology; Volume 2 Classes of Pesticides. Academic Press, Inc., NY. 1991.
94. National Library of Medicine: Hazardous Substances Databank. TOXNET, Medlars Management Section, Bethesda, MD. 1992.
95. Bustos-Obregón E, González-Hormazábal P. Effect of a single dose of malathion on spermatogenesis in mice. Asian J Androl 2003;5:105–7.
96. Krause W, Hamm K, Weissmuller J. The effect of perorally administered DDVP and malathion on spermatogenesis and Leydig cells in the juvenile rat. Andrologia 1975;7:109–16.

97. Krause W. Influence of DDT, DDVP and malathion on FSH, LH and testosterone serum levels and testosterone concentration in testis. Bull Environ Contam Toxicol 1977;18:231–42.

98. Russell L, Ettlin R, Sinha Hikim A, Clegg E. Histological and histopathological evaluation of the testis. Clearwater, Cache River. 1990.

99. Yavasoglu A, Sayim F, Uyanikgil Y, Turgut M, Karabay-Yavasoglu NU. The pyrethroid cypermethrin-induced biochemical and histological alterations in rat liver. J Health Sci 2006;52(6):774–80.

100. Abdel Razik H, Farrag H, Shalby SEM. Comparative histopathological and histochemical studies on IGR, lufenuron and profenofos insecticide albino rats. J Appl Sci Res 2007;3(5):377–86.

101. Afshar S, Farshid AA, Heidari R, Ilkhanipour M. Histopathological changes in the liver and kidney tissues of Wistar albino rat exposed to fenitrothion. Toxicol Industr Health 2008;24(9):581–6.

102. Tos-Luty SD Obuchowska-Przebirowska, Latuszynska J, Tokarska-Rodak M, Haratym-Maj A. Dermal and oral toxicity of malathion in rats. Ann Agri Environ Med 2003;10(1):101–6.

4

Ecotoxicology of Glyphosate and Glyphosate-Based Herbicides - Toxicity to Wildlife and Humans

1. Introduction

The use of agrochemicals is necessary to control pests and increase yields in order to produce adequate food for the global population, estimated at 6.8 billion in 2009 [1], and recently reported to have reached 7 billion [2]. Developing countries, where 1.02 billion people (15 %) are undernourished and 1.3 billion people (19 %) live on an inadequate diet [1], need an adequate food supply. However, the agricultural sector's annual application of over 140 billion kilograms of fertilizers and large amounts of pesticides creates massive sources of diffuse pollution of freshwater ecosystems [3]. In an attempt to increase food production, there is extensive use of herbicides without much regard to the consequences posed to the environment and humans. Glyphosate-based herbicides, which are extensively used in genetically modified glyphosate-resistant plants, are found all over the world [4] and have been reported to occur in various quantities in the aquatic ecosystem, wildlife and humans.

Globally, the presence of pesticides accumulation in both wildlife and humans is on ascendancy, with the health and normal functioning of the endocrine systems being at risk [5-7]. It is believed that the effects of these chemicals on normal functioning of the endocrine system are responsible for a number of developmental anomalies in a wide range of species, from invertebrates to higher mammals [8-11].

The aquatic environment is a receptacle of several undesirable contaminants, including agrochemicals. Therefore, contamination of the aquatic environment by pesticides has become a huge environmental concern worldwide [12]. Glyphosate and glyphosate-based herbicides are among the most widely used class of pesticides. Roundup®, is a major glyphosate-based herbicide used worldwide. Over the years, studies have suggested adverse effects of glyphosate and glyphosate-based compounds on terrestrial and aquatic environments, but recent publications are alluding to the possible effects of glyphosate on mammals, including humans, at different levels of biological organisations as well. In this chapter, the toxicology of glyphosate and glyphosate-based herbicides are explored. We reviewed the impacts of glyphosate and glyphosate-based herbicides on wildlife and humans using measured endpoint effects caused by genotoxicity, cytotoxicity and reproductive toxicity.

1.1. Pesticide as pollutants of freshwater ecosystems

Pesticides are mixtures of chemical substances designed to control, repel, mitigate, kill or regulate the growth of undesirable and nuisance biological organisms [13]. Pests include plant pathogens, weeds, nematodes, molluscs, insects, fish, birds, mammals and microorganisms such as bacteria and viruses. They compete with humans for food, transmit diseases and destroy crops as well as properties [13]. There are various ways of classifying pesticides, with the classification based on the type of pest they control being the most common. For example, insecticides, herbicides, fungicides, nematicides and rodenticides are used to control insects, weeds, fungi, nematodes and rodents, respectively. Furthermore, majority of pesticides are synthetic as they are formulated through industrial processes, while a few are biological as they are derived from natural sources. In addition, broad-spectrum pesticides are applied in controlling a wide range of species but narrow-spectrum pesticides control a small group of pests [13].

Although pesticides are used in agriculture to maintain high production efficiency, they may be environmental hazards and pose risk particularly to non-targeted organisms, and generally to aquatic ecosystems [14, 15]. The potential of a pesticide's risk to an aquatic ecosystem is influenced by its properties, including half-life, mobility and solubility [13]. Microbial activity, drainage pattern, rainfall, treatment surface and application rate can also affect pesticidal activity on a local, regional or global scale [16, 17]. Pesticides get into aquatic systems through processes such as direct applications, surface runoffs, spray drifts, agricultural returns and groundwater intrusions [18]. Pesticides found in urban and agricultural settings in recent times have been implicated in the deaths of many aquatic biota [19].

1.2. Presence of herbicides in freshwater ecosystems

Weeds are unwanted vegetation, which are not planted intentionally, but inadvertently grow in unexpected places. They are usually controlled (i.e., killed or supressed) by the application of a specific herbicide type or class. Classification of herbicides may depend on the criteria used. The two most common criteria employed in herbicide classification are based on time of application and mode of action [20]. Table 1 shows herbicides classification based on time of application and mode of action. Herbicides, which are widely used to control weeds in forestry and agriculture, can reach the aquatic ecosystems by uncontrolled runoff, aerial drift or inadvertent overspray. In some cases, herbicides are directly sprayed at aquatic weeds (e.g. water hyacinth) found on surfaces of water bodies as a control measure. All these impact the aquatic biota.

Table 1. Examples of herbicide classification based on time of application and mode of action (* Note: there may be more than one chemical family for each category of herbicide)

Classification	Chemical family*	Examples
Time of herbicide application		
Pre-emergence: applied to the soil after the crop is planted	Dinitroaniline	Pendulum AquaCap Oryzalin (Surflan AS)
Post-emergence: applied to both crop and weeds after they have germinated and emerged from the soil	Benzoxazole	Acclaim® Extra
Mode of action		
Hormone inhibitors: These herbicides inhibit cell division and growth in the meristem regions (growing points) by mimicking IAA, the natural plant hormone. This interferes with cell wall plasticity and nucleic acid metabolism.	Phenoxycarboxylic acid	2,4-D 2,4-DB 2,4-DP MCPA MCPB
Cell division inhibitors: These herbicides bind to tubulin, the major microtubule protein, to form a herbicide–tubulin complex, leading to a loss of microtubule structure and function. Herbicide-induced microtubule loss may cause cells to neither divide nor elongate, which may be observed as swelling of root tips.	Carbamate	Chlorpropham Propham Carbetamide
Photosynthesis inhibitors: These herbicides inhibit photosynthesis by preventing electron flow, CO_2 fixation and, ATP and NADPH2 production in the photosystem II complex in chloroplasts. Lack of ATP and NADPH2, as well as free radicals destroy cell membranes lead to eventual death of plant.	Triazine	Atrazine, Simazine, Caparol
Lipid synthesis inhibitors: These herbicides inhibit fatty acid and lipid biosynthesis. This causes reduction in cuticular wax development and eventual death of plant.	Thiocarbamate	Cycloate Dimepiperate Pebulate Thiobencarb Triallate

Classification	Chemical family*	Examples
Cell metabolism inhibitors: These herbicides capture electrons from photosystem I, reduce them to form herbicide free radicals, which then destroy cell membranes.	Bipyridylium	Diquat Paraquat Gramoxone
EPSP Synthase Inhibitors: These herbicides inhibit EPSP synthase enzyme, which leads to the depletion of the aromatic amino acids tryptophan, tyrosine and phenylalanine.	Glycines	Glyphosate

The potential of some herbicides to control unwanted vegetation is inherent in their chemical nature, while others have additives to enhance their efficacy. These additives include carriers and adjuvants. In recent years, carriers and adjuvants have been implicated in adding to the toxicity of the active ingredients, and in some cases, have been even more toxic than the active ingredient alone [20]. Prior to the registration of herbicide products for use, not only does the herbicidal properties (Table 2) are assessed, but also the potential effects on humans, animals and environmental safety are assessed. The inherent toxicity of a herbicide, concentration to which an organism is exposed, and duration of exposure determine the extent to which the herbicide can adversely affect an aquatic organism [21]. Herbicides may reach aquatic ecosystems directly by an overhead spray of aquatic weeds, or indirectly through processes such as agricultural runoff, spray drift and leaching [13]. Potential problems associated with herbicide-use include injury to non-target vegetation, injury to crops, residue in soil or water, toxicity to non-target organisms, and concerns for human health and safety [20]. Herbicides can influence the environmental water quality and ecosystem functioning by reducing species diversity, changing community structure, modifying food chains, altering patterns of energy flow and nutrient recycling, as well as reducing resilience of ecosystems [22].

Table 2. Herbicidal properties of herbicides that enhance their efficacy

Herbicidal property	Explanation
Chemical structure	The biologically active portion of a herbicide product is the *active ingredient*. It is the fundamental molecular composition and configuration of the herbicide. The physical and chemical properties of a herbicide can also determine the method of application and use.
Water solubility and polarity	Herbicides that are produced as salts dissolve quite well in water and are usually formulated to be applied in water, while non-polar herbicide sources are not. Water is the main substance used to disperse (spray) herbicides, and hence the water solubility of a herbicide influences the type of product that is formulated, how it is applied and the movement of the herbicide in the soil profile.

Herbicidal property	Explanation
Volatility	Herbicides with a high vapour pressure volatilise easily, while those with a low vapour pressure are relatively non-volatile. The volatility of a herbicide can determine the mode of action and the herbicide's fate in the environment.
Formulations	Commercial herbicide products contain an active ingredient and "inert" ingredients. An "inert" ingredient could be a carrier that is used to dilute and disperse the herbicide (e.g. water, oil, certain types of clay, vermiculite, plant residues, starch polymers, certain dry fertilizers) or an adjuvant (e.g. activator, additive, dispersing agent, emulsifier, spreader, sticker, surfactant, thickener, wetting agent) that enhances the herbicide's performance, handling, or application.

2. Glyphosate and Glyphosate-based Herbicides

Glyphosate (N-(phosphonomethyl) glycine) (Figure 1) and glyphosate-based herbicides are the world's leading post-emergent, organophosphonate systemic, broad-spectrum and non-selective herbicides for the control of annual and perennial weeds [22, 23]. Worldwide, the number one glyphosate-based herbicide used is Roundup®. Other trade names of glyphosate-based herbicides include Roundup Ultra®, Roundup Pro®, Accord®, Honcho®, Pondmaster®, Protocol®, Rascal®, Expedite®, Ranger®, Bronco®, Campain®, Landmaster®, Fallow Master® and Aquamaster® manufactured by Monsanto; Glyphomax®, Glypro® and Rodeo® manufactured by Dow Agrosciences; Glyphosate herbicide manufactured by Du Pont; Silhouette® manufactured by Cenex/Land O'Lakes; Rattler® manufactured by Helena; MirageR® manufactured by Platte; JuryR® manufactured by Riverside/ Terra; and Touchdown® manufactured by Zeneca [24-26].

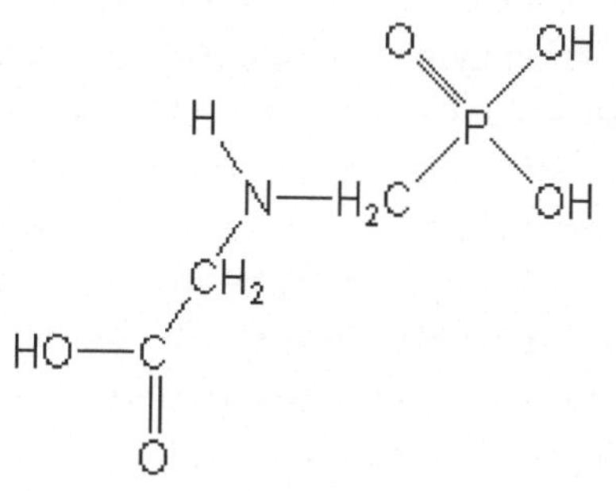

Figure 1. Molecular structure of N-(phosphonomethyl) glycine

Glyphosate has relatively low solubility in water (12 g/L at 25° C and 60 g/L at 100° C), but is insoluble in other solvents [27]. Therefore, commercial formulations of glyphosate are usually in the form of salt to ensure higher solubility yet maintaining the herbicidal properties of the parent compound [22]. Formulations of glyphosate in salt form include monoammonium salt, diammonium salt, isopropylamine salt, potassium salt, sodium salt, and trimethylsulfonium or trimesium salt. Of these, the isopropylamine, sodium, and monoammonium salt forms are commonly used in formulated herbicide products [28]. The isopropylamine salt is the most commonly used in commercial formulated products (e.g. Roundup®). The concentration of glyphosate is commonly expressed as mg a.i./L (active ingredient/Litre) or mg a.e./L (acid equivalents/Litre) [22]. Acid equivalent is the theoretical per cent yield of parent acid from a pesticide active ingredient, which has been formulated as a derivative (usually esters, salts or amines) [29].

2.1. Mode of action of glyphosate

As a systemic herbicide, glyphosate is readily translocated through the phloem to all parts of the plant. Glyphosate molecules are absorbed from the leaf surface into plant cells where they are symplastically translocated to the meristems of growing plants [22]. Glyphosate's phytotoxic symptoms usually start gradually, becoming visible within two to four days in most annual weeds, but may not occur until after seven days in most perennial weeds. Physical phytotoxic symptoms include progress from gradual wilting and chlorosis, to complete browning, total deterioration and finally, death [22]. The primary mode of action of glyphosate is confined to the shikimate pathway aromatic amino acid biosynthesis, a pathway that links primary and secondary metabolisms.

Shikimate (shikimic acid) is an important biochemical intermediary in plants and microorganisms, such as bacteria and fungi. It is a precursor for the aromatic amino acids phenylalanine, tryptophan and tyrosine. Other precursors of the shikimate pathway are indole, indole derivatives (e.g. indole acetic acid), tannins, flavonoids, lignin, many alkaloids, and other aromatic metabolites. The biosynthesis of these essential substances is promoted by enzyme 5-enolpyruvylshikimate-3-phosphate synthase (EPSPS), the target enzyme of glyphosate. This enzyme is one of the seven enzymes that catalyse a series of reactions, which begins with the reaction between shikimate-3-phosphate (S3P) and phosphoenolpyruvate (PEP). The shikimate pathway accounts for about 35 % of the plant mass in dry weight and therefore any interference in the pathway is highly detrimental to the plant. Glyphosate inhibits the activity of EPSPS, preventing the production of chorismate – the last common precursor in the biosynthesis of numerous aromatic compounds in bacteria, fungi and plants. This causes a deficiency in the production of the essential substances needed by the organisms to survive and propagate [22, 30]. The pathway is absent in animals, which may account for the low toxicity of glyphosate to animals.

However, acute effects in animals, following intraperitoneal administration of high glyphosate doses, suggest altered mitochondrial activity, possibly due to uncoupling of oxidative phosphorylation during cellular respiration [27]. In summary, glyphosate ultimately interrupts various biochemical processes, including nucleic acid synthesis, protein synthesis, photosynthesis and respiration, which are essential life processes of living things.

2.2. Environmental fate of glyphosate

Glyphosate has a strong soil adsorption capacity, which limits its movement in the environment. The average half-life of glyphosate in soil is two months, but can range from weeks to years [24]. Glyphosate in freshwater ecosystems has an average half-life of two to ten weeks [24]. The rate of degradation in water is generally slower than in most soils because of fewer microorganisms in water than in soils [31]. When glyphosate undergoes degradation, it produces aminomethylphosphonic acid (AMPA) and carbon dioxide [32], both of which reduce pH when dissolved in water. However, pH is known to affect the stability of glyphosate in water. For instance, glyphosate did not undergo hydrolysis in buffered solution with a pH of 3, 6 or 9 at 35° C, while insignificant photodegradation has been recorded under natural light in pH 5, 7 and 9 buffered solutions [28]. In freshwater ecosystems, glyphosate dissipates through degradation, dilution, and adsorption on organic substances, inorganic clays and the sediment (the major sink for glyphosate in water bodies) [24, 31]. With its long half-life and its ability to cause death of organisms in aquatic ecosystems, it is recommended that glyphosate should be used as an aquatic herbicide to treat only one-third to half a water body at any one time [24].

2.3. Toxicology of glyphosate and its effects on aquatic organisms

In recent years, the exposure of non-target aquatic organisms to glyphosate-based herbicides has aroused great concern globally because of high water solubility and the extensive use of glyphosate-based herbicides [25]. In this regard, polyoxyethylene amine (POEA), a surfactant, has been implicated as being the main cause of the relatively high toxicity of Roundup® to several freshwater invertebrates and fishes [25, 33]. Technical grade glyphosate is slightly to very slightly toxic, with reported LC50 values of greater than 55 mg/L and a 21 d NOEC (no observed effect concentration) value of 100 mg/L [25, 33].

Conversely, formulations of glyphosate are moderately to very slightly toxic with 2 d EC50 values of 5.3-5600 mg/L and 21 d MATC values of 1.4-4.9 mg/L reported [27]. The LC50 values also determine which glyphosate formulation can be applied in aquatic ecosystems. It should be noted that high LC50 value of a chemical substance to an organism implies low toxicity of that particular chemical substance to that particular organism, and the reverse is also true. For instance, Rodeo® has relatively high LC50s (>900 mg/L) for aquatic species and is permitted for use in aquatic ecosystems, while Touchdown -4LC® and Bronco® have low LC50 values for aquatic species (<13 mg/L), and are not registered for aquatic use [24]. Similarly, Roundup® is not registered for use in aquatic ecosystems in the United States because its 96 h LC50 value for *Daphnia* is 25.5 mg/L, while that of glyphosate alone is 962 mg/L [24].

In recent years, glyphosate has been found in surface waters long after it has been used to control aquatic weeds, although it is generally regarded as having a low potential for contaminating surface waters [13, 34]. In fact, its mode of action was designed to affect only plants [30], but various studies in recent years have reported adverse impact on non-target animals [25, 33, 35].

2.4. Wildlife toxicology of glyphosate and glyphosate-based herbicides

Wildlife ecology is the application of ecological principles to the study of wildlife species. The term wildlife, however, lacks a universally accepted definition, and its common use changed during the 1900s in association with development of the profession of wildlife management. Historically, wildlife management focused on hunted or harvested birds and mammals that were collectively referred to as game species. However, current description of wildlife ecology includes different levels of biological organisation as well as individual organisms within a population and their interactions with the environment [36]. The effects of environmental contaminants on the health and persistence of wildlife populations have been a concern of environmentalists for many decades [36]. This led to increased interest in the study of exposure of wildlife to environmental contaminants, and hence wildlife toxicology.

A growing collection of wildlife toxicological studies examining diverse wildlife species demonstrates that exposure to environmental contaminants over the years is the cause of increasing disappearance of certain species. Such a threat to global biodiversity usually starts with developments of abnormalities in contaminated organisms. These may include the disruption of genetic material, cell integrity and major but subtle birth defects in individual species. This is against the belief, held for a long time since introduction of glyphosate as a commercial herbicide in the 1970s, that glyphosate is non-toxic to wildlife and humans. In fact, US EPA has classified glyphosate as a "Group E carcinogen", implying it is "non-carcinogenicity for humans" [37]. The main reason for this has been attributed to the fact that the herbicidal activity of glyphosate targets specifically inhibition of the shikimate pathway, which is only present in plants and microorganisms, but conspicuously absent in animals and humans [38]. However, recent studies seem to suggest that glyphosate has genotoxicity, cytotoxicity and reproductive toxicity in wildlife and humans [39-40].

Genotoxicity is a term used to describe the destructive effect by toxic agents referred to as genotoxins on a cell's genetic material (DNA, RNA). Genotoxins include both radiation and chemical genotoxins. There are three primary effects that genotoxins can have on organisms by affecting their genetic information depending on the type of genotoxin. These include cancer-causing agents (carcinogens), mutation-causing agents (mutagens) or birth defect-causing agents (teratogens) [41]. Conversely, cytotoxicity is the term used to describe destructive effect of cells by agents referred to as cytotoxins. Cytotoxic cells may undergo necrosis, in which they lose membrane integrity and die rapidly as a result of cell lysis; they can stop growing and dividing; or they can activate a genetic program of controlled cell death, also referred to as apoptosis [41]. Reproductive toxicity refers to the toxic effects of a substance on the reproductive ability of an organism and the development of its offspring. The Globally Harmonized System (GHS) defines reproductive toxicity as adverse effects of chemical substances on sexual function and fertility in adult males and females, as well as developmental toxicity in the offspring [42]. Developmental toxicity refers to adverse effects induced during pregnancy, or as a result of parental exposure to a chemical substance manifested at any point in the life span of the organism [42]. Thus, reproductive toxicity may be grouped into two main categories: adverse effects on reproductive ability or capacity, and adverse effects on development of the offspring. Developmental toxicity and reproductive toxicity tests are usually performed with female and male animals, embryonic stem cells or whole embryos. In the following sections, we review a few of the many reported studies on adverse effects caused by glyphosate and glyphosate-based herbicides as a result of genotoxicity, cytotoxicity and reproductive toxicity in wildlife as well as in humans.

3. Toxicity of Glyphosate and Glyphosate-based Herbicides to Wildlife

It has been argued that the seasonality of crop cultivation and subsequent seasonal application of herbicides means that their presence in aquatic ecosystems is typically periodic. More so, episodic exposures are often for a short period, which is followed by long periods of non-application that has the capacity to "water down" the exposure effects. Therefore, the assessment of genotoxic endpoints in fish after removal of the contamination source is essential to understanding the dynamics of herbicide genotoxicity and risk posed by these agrochemicals. In order to test this assertion, Guilherme et al. [12] investigated the genotoxic potential of Roundup® to European eel (*Anguilla anguilla*) [12]. *A. anguilla* was exposed to 116 μg/L Roundup® for 3 days, and allowed to recover for 7 ,1 and 14 days after exposure has ceased. A comet assay was applied to identify DNA damage in blood cells during both exposure and post-exposure periods. Guilherme et al. [12] reported that Roundup® has genotoxic potential and exhibited oxidative DNA damage (pyrimidine bases oxidation), although a recovery was evident when considering non-specific DNA damage on day 14 post exposure. Roundup® was able to induce a late oxidative DNA damage (day 14). In the report, Guilherme et al. suggested that blood cells of *A. anguilla* exposed to Roundup® appeared to be more successful in repairing damage with a non-specific cause than that associated with base oxidation [12]. Overall, the findings of this study reinforce the hypothesis that Roundup® as an agrochemical may cause long-lasting damage to fish due to genotoxicity.

In another study, Vera-Candioti et al. evaluated the genotoxicity and cytotoxicity of glyphosate-based formulations Panzer and Credit® after exposure to *Cnesterodon decemmaculatus* (Jenyns, 1842) (Pisces, Poeciliidae) under laboratory conditions

[43]. They used micronuclei induction (MN) and alterations in the erythrocytes to erythroblasts ratio for genotoxicity and cytotoxicity as endpoints, respectively. They reported that both 3.9 and 7.8 mg/L of Panzer increased MN frequency at 48 and 96 h of treatment. Similarly, *C. decemmaculatus* exposed to tested concentration of Credit® had increased frequency of MN over control group after 96 h, but not after 48 h. However, they reported that both Panzer and Credit® treatment did not cause cytotoxicity in fish, regardless of the concentration and sampling time. In their conclusion, Vera-Candioti et al. suggested that Panzer and Credit® should be considered as glyphosate-based commercial formulations with genotoxic but not cytotoxic effect properties [43].

Low levels of glyphosate-based herbicide have also been reported to cause adverse effects on reproduction (fecundity) and clutch size of aquatic invertebrates. In a study, Cuhra et al. reported that *Daphnia magna* exposed to 0.45 mg/L Roundup® significantly decreased fecundity compared to the control group [44]. They also reported that animals exposed to 1.35 mg/L Roundup® reached reproductive age, but almost all eggs and developing embryos were aborted and did not hatch, whereas animals exposed to 4.05 mg/L Roundup® died before reaching maturation. However, fecundity in *Daphnia magna* exposed to Roundup® concentrations of 0.05 and 0.15 mg/L was not significantly affected. The abortion rates for animals exposed to Roundup® concentrations of 0.15 ,0.05 and 0.45 mg/L were not significantly different from those of the control group. On the other hand, the abortion rates for animals exposed to 1.35 mg/L of Roundup® were significantly higher than control group, reaching nearly 100 %. Cuhra et al. also reported that the size of first clutch (brood) juveniles born from groups exposed to 0.15 ,0.05 and 0.45 mg/L Roundup® were not significantly different from the control group [44]. However the size of second clutch juveniles from animals exposed to 0.05, 0.15 and 0.45 mg/L Roundup® were significantly smaller than those of the control group. The authors contextualized their findings by comparing them to the general US EPA environmental guideline limit for glyphosate and the California State's specific EPA environmental guideline limit for glyphosate, which were 0.7 and 1.0 mg/L, respectively. These limits for glyphosate in surface waters are between the 0.45 and 1.35 mg/L concentrations Cuhra et al. used in their study [44]. They concluded that since *D. magna* showed complete reproductive failure and aborted all eggs in early-to-late stages of embryonic development when exposed to 1.35 mg/L suggests that the US EPA and the California State's environmental guidelines may not be sufficiently restrictive to ensure viable populations of *D. magna* and other aquatic invertebrates.

Studies with rats seemed to suggest that the respiratory and hepatic systems as well as reproductive functions including sperm production or libido, and even foetal development can be altered by exposure to Roundup®. This may be attributed to the fact that xenobiotics in the aquatic ecosystem may lead to endocrine disruption at a reproductive and more specifically testicular level in young as well as adult mammals [45]. For example, different forms of testicular dysgenesis (abnormal development and growth of the testicles) have been linked to xenobiotics in aquatic ecosystems. These include decrease in sperm quantity and quality, increase in congenital malformations such as cryptorchidism (the absence of one or both testes from the scrotum) and hypospadias (abnormally placed urinary meatus), and preoccupying increase of testicular cancer incidences [46-49].

In a recent study, Prasad et al [50] investigated the genotoxic effects of glyphosate in the cells of Swiss Albino mice by measuring chromosomal aberrations (CAs) and micronuclei (MN) in bone marrow cells after exposure for 24, 48 and 72 h. Glyphosate treatment group mice were exposed by giving them a single dose of glyphosate intraperitoneally (i.p) at a concentration of 25 and 50 mg/kg b.wt. Simultaneously, positive control group mice were injected i.p. benzo(a)pyrene (100 mg/kg b.wt, once only), while control (vehicle) group mice were injected i.p. dimethyl sulfoxide (0.2 mL). Mice from all the groups were sacrificed at sampling times of 24, 48, and 72 h and their bone marrow analysed for cytogenetic and chromosomal damage. They reported that CAs and MN induction increased significantly in glyphosate treatment groups at both given dosages and time compared with the vehicle control group ($P < 0.05$). They also reported that glyphosate caused cytotoxic effects in the mice by significant decrease in mitotic index (MI). Based on their findings, Prasad et al [50] concluded that glyphosate is clastogenic and cytotoxic to mouse bone marrow.

In another study by Clair et al., mature rat fresh testicular cells were exposed to glyphosate and its formulation, Roundup®, from 1 to 10000 ppm [51]. This is the concentration range reported in some human urine and the environment, as well as in agricultural application levels. They found that Leydig cells got damaged from 1 to 48 h of Roundup® exposure, while other cells where damaged within 48-24 h; all mainly caused by necrosis. By contrast, glyphosate alone was toxic on Sertoli cells and later induced apoptosis at higher doses in germ cells and in Sertoli/germ cells co-cultures. At lower concentrations (i.e. 1 ppm), Roundup® and glyphosate were found to impact the endocrine system as they caused %35 decrease in testosterone, but only a high contamination appears to induce an acute rat testicular toxicity.

The effects of glyphosate on wildlife is not all that gloomy. At least, one study has reported this. Acacia et al. [52] studied the effects of glyphosate (as an active ingredient) and Roundup® (as a formulation) on oyster gametes and embryos to find a possible link between genotoxicity and reproduction/developmental impairment. They wanted to explore the impact of chemical genotoxicity on population dynamics of oysters since glyphosate is frequently found in oyster production areas, among other herbicides. Considering that oyster's gametes and embryos are in direct contact with the surrounding waters because its mode of reproduction is external, the presence of these agrochemicals does not only pose risk to oysters but also other aquatic organisms. In their study, Akcha et al. [52] exposed oyster spermatozoa and embryos to 0.5; 1.0; 1.5; 2.5; 5.0 μg active substance/L of both glyphosate and Roundup®. They reported that glyphosate and Roundup® had no effect on the oyster development at the concentrations tested. Their spermiotoxicity study also showed neither glyphosate nor Roundup® to be cytotoxic for oyster spermatozoa. It should be noted that although these findings by Akcha et al. [52] showed no negative effect on sperm function, the possible impact on fertilization rate and the consequences of the transmission of damaged DNA for oyster development and physiological performances was not investigated. More importantly, Akcha et al. [52] findings suggest that proper monitoring of usage of agrochemicals has the potential of minimising their toxic effects in the environment.

4. Toxicity of Glyphosate and Glyphosate-based Herbicides to Humans

Since the commercial introduction of glyphosate as a herbicide, its health effects have been studied intensely with the general conclusion that it is safe for humans [27, 42, 53]. However, recent published studies indicate that occupational exposure of humans to the herbicide is associated with increased cancer risks [54-55]. Also during the course of production of the herbicide, humans may come into dermal and/or inhalative contact with it. Although the shikimic pathway through which glyphosate disrupts plants biochemical activities is not found in animals, various studies seem to suggest the agrochemical may affect other pathways. The mitochondria and cytochrome P450 (CYP) pathways are thought to include possible candidate sites of action in animals, although there is no solid evidence to support such claims.

Mesnage et al. [56] investigated potential toxicity of 9 glyphosate-based formulations as well as technical grade glyphosate and polyethoxylated tallowamine POE-15 (the major adjuvant used in glyphosate-based formulations) to human cells, including hepatic (HepG2), embryonic (HEK293) and placental (JEG3) cell lines, after 24 h exposures [56]. They measured mitochondrial activities, membrane degradations and caspases 3/7 activities as endpoints. The authors reported that all formulations were more toxic than glyphosate, but POE-15 was found to be the most toxic against human cells, even if others were not excluded. The toxicity effect began with negative dose-dependent effects on cellular respiration and membrane integrity between 1 and 3 mg/L at environmental/occupational doses. They reported that POE-15 induced necrosis when its first micellisation process occurred, while glyphosate promoted endocrine disrupting effects after entering cells. The findings of Mesnage et al. challenged the establishment of guidance values such as the acceptable daily intake (ADI) of glyphosate since these are mostly based on a long term in vivo test of glyphosate alone [56]. The authors suggested that it is imperative to assess whole formulations of pesticides as mixtures with adjuvants that could change their toxicity in pesticide toxicity investigations.

Another study by Samsel and Seneff [57] asserted that glyphosate is minimally toxic to humans since residues are found in food stuff, including sugar, corn, soy and wheat. In their opinion, the disruption of cytochrome P450 (CYP) enzymes activities by glyphosate is an overlooked component of its toxicity to mammals. In a recent study, they reported that glyphosate interferences with cytochrome P450 (CYP) enzymes and acts synergistically to disrupt the biosynthesis of aromatic amino acids by gut bacteria, as well as impairment in serum sulphate transport. According to the authors, one of the many crucial functions of CYP enzymes is detoxification of xenobiotics. Therefore, by disrupting CYP enzymes activities, glyphosate enhances the damaging effects of other food borne chemical residues and environmental toxins. This adversely affects the body though the impact is subtle and manifests slowly over time as inflammation damages cellular systems throughout the body. Ultimately, these result in diseases and conditions such as gastrointestinal disorders, obesity, diabetes, heart disease, depression, autism, infertility, cancer and Alzheimer's disease; mostly associated with a Western diet. Based on their study outcome, Samsel and Seneff described glyphosate as "textbook example" of exogenous semiotic entropy: the disruption of homeostasis by environmental toxins".

Further recent studies have also suggested glyphosate to be an endocrine disrupting chemical (EDC), which has the potential to cause adverse health effects in humans [58]. In a study to evaluate the EDC properties of glyphosate in humans, Thongprakaisang et al [58] investigated the effects of technical grade glyphosate on estrogen receptors (ERs) mediated transcriptional activity and their expressions [58]. They reported that the proliferative concentrations of glyphosate, which caused the activation of estrogen response element (ERE) transcription activity, were 5-13 fold more than the control in T47D-KBluc cells. However, the activation was inhibited by an estrogen antagonist, ICI 182780, which implied that the estrogenic activity of glyphosate was mediated via ERs. The findings of this investigation suggest that low and environmentally relevant concentration of glyphosate can disrupt the hormonal systems of humans. Furthermore, the effects demonstrate that glyphosate is or could act as a "xenoestrogen" and may be capable of inducing EREs in a manner slightly weaker but functionally similar to Estradiol (E2), the most potent human estrogen.

In another study to investigate the xenobiotic toxicity of glyphosate, Gasnier et al. exposed the human liver HepG2 cells to four different formulations and to glyphosate, and measured the cytotoxicity, genotoxicity, anti-estrogenic (on estrogen receptors (ERα) and (ERβ)) and anti-androgenic effects (on androgen receptor (AR)), as well as checked androgen to estrogen conversion by aromatase activity and mRNA [59]. They reported that all parameters were disrupted at sub-agricultural concentrations with all formulations within 24 h, with the effects more dependent on the formulation than on the glyphosate concentration. They also stated that concentration levels above 0.5 mg/L of the most active formulation

(R400) caused a human cell endocrine disruption on the androgen receptor in MDA-MB453-kb2 cells, while concentration levels above 2 mg/L inhibited transcriptional activities on both ERs on the HepG2. The authors also reported that concentration levels above 10 mg/L disrupted aromatase transcription and activity, while cytotoxic effects started at concentration levels above 10 mg/L and DNA damages (genotoxic effects) at 5 mg/L.

Richard et al also investigated the effects of glyphosate and Roundup® on human placental JEG3 cells within 18 h with concentrations lower than those found with agricultural use [60]. They stated that both chemical substances were toxic to the human placental JEG3 cells and this effect increases with concentration and time but Roundup® was found to be more toxic than glyphosate. They also tested the effects of both chemical substances at lower environmentally nontoxic concentrations on aromatase, the enzyme responsible for estrogen synthesis. They reported that Roundup® disrupted aromatase activity and mRNA levels and interacted with the active site of the purified enzyme, but the effects of glyphosate were facilitated by adjuvants in microsomes or in cell culture. Based on their findings, Richard® et al suggested that glyphosate and Roundup® can induce endocrine and toxic effects in humans and other mammals.

In a separate study to ascertain whether glyphosate exposure may cause DNA damage and cancer in humans, Koller et al. investigated exposure of workers via inhalation to technical glyphosate and Roundup UltraMax, glyphosate-based herbicide [61]. They reported that the cytotoxic and genotoxic properties of glyphosate in the workers' buccal epithelial cell line (TR146) induced acute cytotoxic effects at concentration 40 mg/L after 20 min, which were due to membrane damage and impairment of mitochondrial functions. Similarly, they stated that Roundup UltraMax induced release of extracellular lactate dehydrogenase at concentrations 80 mg/L, which indicates membrane damage. In their study, Koller et al. showed that both glyphosate and Roundup UltraMax induced DNA migration in single-cell gel electrophoresis assays at concentrations 20 mg/L [61]. The authors again showed that the frequencies of micronuclei and nuclear buds were elevated after 20 min exposure to 10-20 mg/L, as well as increase of nuclear aberrations that reflected DNA damage. However, nucleoplasmatic bridges were only enhanced by glyphosate at the highest dose (20 mg/L). The authors concluded that their findings suggest that inhalation of glyphosate and glyphosate-based herbicides by humans may cause DNA damage in exposed individuals since they found genotoxic effects after short exposure to concentrations that correspond to a 450-fold dilution of spraying used in agriculture [61].

5. Conclusion

In this chapter, we discussed the perceived "friendly" nature of glyphosate and glyphosate-based herbicides and their apparent non-toxicity to both wildlife and humans. This notion has been in existence ever since glyphosate was first introduced as a commercial herbicide in the 1970s largely because the herbicidal activity of glyphosate targets the inhibition of the shikimate pathway in particular, which is only present in plants and microorganisms. Thus glyphosate is classified by the US EPA as a Group E carcinogen, and therefore is non-carcinogenic to humans. However, current investigations involving glyphosate exposure to wildlife and humans show adverse effects resulting from genotoxicity, cytotoxicity and reproductive toxicity. These have been reviewed with examples in this chapter. Furthermore, we have demonstrated that the herbicide has an endocrine impact at very low environmentally relevant concentrations. This review of glyphosate herbicides needs to be taken seriously since the use of glyphosate-contaminated plant products as dietary supplements may pose a risk of cancer in humans because of their potential additive estrogenicity. In addition, the extensive use of glyphosate-based herbicides in genetically modified glyphosate-resistant plants grown for food and feed should be of grave concern since they can be sources of genotoxicity, cytotoxicity, and reproductive toxicity in wildlife and humans.

References

1. Jurado, A.S., Fernandes, M.A.S., Videira, R.A., Peixoto, F.P. and Vicente, J.A.F. (2011). Herbicides: The Face and the Reverse of the Coin. An in vitro Approach to the Toxicity of Herbicides in Non-Target Organisms, Herbicides and Environment, Andreas Kortekamp (Ed.), ISBN: 978-953-307-476-4, InTech. Accessed September 2013 at: http://www.intechopen.com/articles/show/title/herbicides-the-face-and-the-reverse-of-the-coin-an-in-vitro-approach-to-the-toxicity-of-herbicides-i.
2. PRB (Population Reference Bureau) (2012). 2012 World population data sheet, published by the Population Reference Bureau, 1875 Connecticut Ave., NW, Suite 520, Washington, DC 20009 USA. Accessed December 2012 at: http://www.prb.org/pdf12/2012-population-data-sheet_eng.pdf.
3. Schwarenbach, R.P., Escher, B.I., Fenner, K., Hoffstetter, T.B., Johnson, C.A., Von Gunten, U. and Wehrli, B. (2006). The challenge of micropollutants in aquatic systems. Science, 313, 1072-1077.
4. Acquavella, J., Alexander, B., Mandel, J., Gustin, C., Baker, B., Chapman, P., Bleeke, M. (2004). Glyphosate biomonitoring for farmers and their families: results from the farm family exposure study. Environmental Health Perspective, 112, 321-326.
5. LeBlanc, G.A. (2007). Crustacean endocrine toxicology: a review. Ecotoxicology, 16, 61-81.
6. Correia, T.G., Narcizo, A.M., Bianchini, A. and Moreira, R.G. (2010). Aluminum as an endocrine disruptor in female Nile tilapia (Oreochromis niloticus). Comparative Biochemistry and Physiology, Part C, 151, 461-466.
7. Pedersen, M., Halldorsson, T.I., Mathiesen, L., Mose, T., Brouwer, A., Hedegaard, M., Steffen, L., Kleinjans, J.C.S., Besselink, H. and Knudsen, L.E. (2010). Dioxin-like exposures and effects on estrogenic and androgenic exposures and micronuclei frequency in mother–newborn pairs. Environment International, 36, 344-351.

8. Abel P.D. (2002). Water Pollution Biology (2nd edn.), Taylor and Francis Ltd, London.

9. London, L., Dalvie, M.A. and Cairncross, E. (2005). Approaches for regulating water in South Africa for the presence of pesticides. Water SA, 31(1), 53-59.

10. Mihaich, E.M., Friederich, U., Caspers, N., Hall, A.T., Klecka, G.M., Dimond, S.S., Staples, C.A., Ortego, L.S. and Hentges, S.G. (2009). Acute and chronic toxicity testing of bisphenol A with aquatic invertebrates and plants. Ecotoxicology and Environmental Safety, 72, 1392-1399.

11. Benstead, R.S., Baynes, A., Casey, D., Routledge, E.J., Jobling, S. (2011). 17 β-Oestradiol may prolong reproduction in seasonally breeding freshwater gastropod molluscs. Aquatic Toxicology, 101, 326-334.

12. Guilherme, S., Santos, M.A., Gaivão, I., Pacheco, M. (2014). Are DNA-damaging effects induced by herbicide formulations (Roundup® and Garlon®) in fish transient and reversible upon cessation of exposure? Aquatic Toxicology, 155, 213-221.

13. Mensah, P.K., Palmer, C.G., Muller, W.J. (2014). Lethal and sublethal effects of pesticides on aquatic organisms: the case of a freshwater shrimp exposure to Roundup®. In: Larramendy, M.L. and Soloneski, S. (Eds), Pesticides: Toxic Aspects, InTech Publications, Rijeka, Croatia, pp. 163-185.

14. Lipika, P. and Patra, A.K. (2006). Haemoatopoietic alterations induced by carbaryl in Clarias batrachus (LINN). Journal of applied Sciences and Environmental Management, 10 (3), 5-7.

15. Boran, M., Altinok, I., Capkin, E., Karacam, H. and Bcer, V. (2007). Acute toxicity of carbaryl, methiocarb and carbosulfan to the rainbow trout (Oncorhynchus mykiss) and guppy (Poecilia reticulate). Turkish Journal of Veterinary and Animal Science, 31.

16. Storrs, S. I. and Kiesiecker (2004). Survivorship patterns of larval amphibians exposed to low concentrations of atrazine. Environmental Health Perspectives, 112 (10), 1054-1057.

17. Maharaj S. (2005). Modelling the Behaviour and Fate of Priority Pesticides in South Africa. MSc Thesis, Department of Earth Sciences, University of Western Cape, South Africa.

18. Scholz, N.L., Incardona, J.P., Baldwin, D.H., Berejikan, B.A., Dittman, A.H., Feist, B.E., Jordan, C. (2003). Evaluating the sublethal impacts of current use pesticides on the environmental health of salmonids in Columbia River Basin. Bonneville Power Administration FY 2003 Provincial Project Review, 1-41

19. Khan, M.Z., Tabassum, R., Naqvi, S.N.H., Shah, E.Z., Tabassum, F., Ahmad, I., Fatima, F., Khan, M.F. (2003). Effect of cypermethrin and permethrin on cholinester ase activity and protein contents in Rana tigrina (Amphibia), Turkey Journal of Zoology, 27, 243-246.

20. Radosevich, S.R., Holt J.S., Ghersa C.M. (2007). Ecology of Weeds and Invasive Plants: Relationship to Agriculture and Natural Resource Management (3rd edn.), Wiley-Interscience, Hoboken, USA.
21. Wilson, C. (2009). Aquatic toxicology notes: predicting the fate and effects of aquatic and ditchbank herbicides. Soil and Water Science Department, Institute of Food and Agricultural Sciences, University of Florida, Gainesville, USA.
22. Pérez, G.L., Vera, M.S., Miranda, L. (2011). Effects of Herbicide Glyphosate and Glyphosate-Based Formulations on Aquatic Ecosystems. In: Andreas Kortekamp (Ed.), Herbicides and Environment. InTech Publications, Rijeka, Croatia.
23. Mink, P.J., Mandel, J.S., Sceurman, B.K., Lundin, J.I. (2012). Epidemiologic studies of glyphosate and cancer: A review. Regulatory Toxicology and Pharmacology, 63, 440-452.
24. Tu, M., Hurd, C., Randall, J.M. (2001).Weed Control Methods Handbook: Tools and Techniques for Use in Natural Areas, The Nature Conservancy. Accessed July 2009 at: http://tncweeds.ucdavis.edu.
25. Tsui, M.T.K., Chu, L.M. (2003). Aquatic toxicity of glyphosate-based formula tions: comparison between different organisms and the effects of environmental factors. Chemosphere, 52, 1189-1197.
26. Zhou, D-M., Wang, Y-J., Cang, L., Hao, X-Z., Luo, X-S. (2004). Adsorption and cosorption of cadmium and glyphosate on two soils with different characteristics. Chemosphere, 57, 1237-1244.
27. WHO (World Health Organization) (1994). Environmental Health Criteria 159 – Glyphosate, International Programme on Chemical Safety. World Health Organization, Geneva, Switzerland.
28. Miller, A., Gervais, J.A., Luukinen, B., Buhl, K., Stone, D. (2010). Glyphosate Technical Fact Sheet; National Pesticide Information Center, Oregon State University Extension Services. Accessed January 2012 at: http://npic.orst.edu/factsheets/glyphotech.html.
29. Nordby, D.E., Hager, A.G. (2011). Herbicide formulations and calculations: active ingredient or acid equivalent, a Weed Fact sheet. Integrated Pest Management Handbook, University of Illinois, USA. Accessed 4 November 2011 at: http://ipm.illinois.edu/weeds/aeai.pdf.
30. Stenersen, J. (2004). Chemical Pesticides: Mode of Action and Toxicology. CRC Press, Boca Raton, Florida, USA, 296 pp.
31. Schuette, J. (1998). Environmental fate of glyphosate, Environmental Monitoring & Pest Management, Department of Pesticide Regulation, Sacramento, CA 95824-5624, USA.
32. Meyer, M.T., Loftin, K.A., Lee, E.A., Hinshaw, G.H., Dietze, J.E., Scribner, E.A. (2009). Determination of Glyphosate, its Degradation Product Aminomethylphosphonic Acid, and Glufosinate, in Water by Isotope Dilution and Online Solid-Phase Extraction and Liquid Chromatography/Tandem Mass Spectrometry. U.S. Geological Survey Techniques and Methods, book 5, chap. A10, 32p.

33. Giesy, J.P., Dobson, S., Solomon, K.R. (2000). Ecotoxicological risk assessment for Roundup herbicide. Review of Environmental Contamination and Toxicology, 167, 35-120.

34. Glusczak, L., Miron, D.S., Moraes, B.S., Simões, R.R., Schetinger, M.R.C., Morsch, V.M., Loro, V.L. (2007). Acute effects of glyphosate herbicide on metabolic and enzymatic parameters of silver catfish (Rhamdia quelen). Comparative Biochemistry and Physiology-Part C, 146, 519-524.

35. El-Shebly, A.A., El-Kady, M.A.H. (2008). Effects of glyphosate herbicide on serum growth hormone (GH) levels and muscle protein content in Nile Tilapia (Oreochromis niloticus L.). Research Journal of Fisheries and Hydrobiology, 3(2), 84-88.

36. vom Saal, F.S., Guillette, J.R. Jr., Myers, J.P., Swan, S. (2008). Endocrine disruptors: Effects in wildlife and laboratory animals. In: Encyclopaedia of Ecology, Jorgensen, S.E., Fath, B.D. (Eds). Pp 1261-1264.

37. US EPA (1993). United States Environmental Protection Agency. Reregistration eligibility decision (RED): glyphosate. EPA 738-R-93-014. Office of Prevention, Pesticides, and Toxic Substances. Washington, DC: US EPA.

38. Solomon, K., Anad, A., Carrasquilla, G., Cerdeira, A., Marshall, E., Sanin, L. (2007). Coca and poppy eradication in Colombia: Environmental and human health assessment of aerially applied glyphosate. Reviews of Environmental Contamination and Toxicology, 190, 43-125.

39. EC (European Commission) (2002). Review report for the active substance glyphosate, Directive 6511/VI/99. January 21, 2002. Directorate E – Food Safety: plant health, animal health and welfare, international questions, E1-Plant health.

40. WHO/FAO (World Health Organization/Food and Agriculture Organization of the United Nations) (2004). Pesticides Residues in Food, Report of the Joint Meeting of the FAO Panel of Experts on Pesticide Residues in Food and the Environment and the WHO Core Assessment Group on Pesticide Residues (JMPR). Rome, Italy, 20-29 September 2004. FAO Plant Production and Protection Paper 178. Rome, Italy.

41. Shah, S.U. (2012). Importance of genotoxicity and S2A guidelines for genotoxicity testing for pharmaceuticals. IOSR Journal of Pharmacy and Biological Sciences, 1(2), 43-54.

42. UNECE (United Nations Economic Commission for Europe) (2013). Globally Harmonized System of Classification and Labelling of Chemicals (GHS), (5th revised edn.), Part 3: Health Hazards. GHS (Rev.5). Accessed January 2015 at: http://www.unece.org/ru/trans/danger/publi/ghs/ghs_rev05/05files_r.html.

43. Vera-Candioti, J., Soloneski, S., Larramendy, M.L. (2013). Evaluation of the genotoxic and cytotoxic effects of glyphosate-based herbicides in the ten spotted live-bearer fish Cnesterodon decemmaculatus (Jenyns, 1842). Ecotoxicology and Environmental Safety, 89, 166-173.

44. Cuhra, M., Traavik, T., Bøhn, T. (2013). Clone- and age-dependent toxicity of a glyphosate commercial formulation and its active ingredient in Daphnia magna. Ecotoxicology, 22, 251-262.
45. Anway, M.D., Memon, M.A., Uzumcu, M., Skinner, M.K. (2006). Transgenerational effect of the endocrine disruptor vinclozolin on male spermatogenesis. Journal of Andrology, 27, 868-879.
46. Carlsen, E., Giwercman, A., Keiding, N., Skakkebaek, N.E. (1992). Evidence for decreasing quality of semen during past 50 years. BMJ 305, 609-613.
47. Auger, J., Kunstmann, J.M., Czyglik, F., Jouannet, P. (1995). Decline in semen quality among fertile men in Paris during the past 20 years. New England Journal of Medicine, 332, 281-285.
48. Bergstrom, R., Adami, H.O., Mohner, M., Zatonski, W., Storm, H., Ekbom, A., Tretli, S., Teppo, L., Akre, O., Hakulinen, T. (1996). Increase in testicular cancer incidence in six European countries: a birth cohort phenomenon. Journal of the National Cancer Institute, 88, 727-733.
49. Toppari, J., Virtanen, H.E., Main, K.M., Skakkebaek, N.E. (2010). Cryptorchidism and hypospadias as a sign of testicular dysgenesis syndrome (TDS): environmental connection. Birth Defects Res. A Clin. Mol. Teratol. 88, 910-919.
50. Prasad, S., Srivastava, S., Singh, M., Shukla, Y. (2009). Clastogenic effects of glyphosate in bone marrow cells of Swiss albino mice. Journal of toxicology, 2009. 6 pages.
51. Clair, É., Mesnage, R., Travert, C., Séralini, G.É. (2012). A glyphosate-based herbicide induces necrosis and apoptosis in mature rat testicular cells in vitro, and testosterone decrease at lower levels. Toxicology in vitro, 26(2), 269-279.
52. Akcha, F., Spagnol, C. and Rouxel, J. (2012). Genotoxicity of diuron and glyphosate in oyster spermatozoa and embryos. Aquatic Toxicology, 106-107, 104-113.
53. Williams, G.M., Kroes, R., Munro, I.C. (2000). Safety evaluation and risk assessment of the herbicide Roundup and its active ingredient, glyphosate, for humans. Regulatory Toxicology and Pharmacology, 31, 117-165.
54. Eriksson, M., Hardell, L., Carlberg, M., Akerman, M. (2008). Pesticide exposure as risk factor for non-Hodgkin lymphoma including histopathological subgroup analysis. International Journal of Cancer, 123, 1657-1663.
55. Bolognesi, C., Carrasquilla, G., Volpi, S., Solomon, K.R., Marshall, E.J. (2009). Biomonitoring of genotoxic risk in agricultural workers from five Colombian regions: association to occupational exposure to glyphosate. Journal of Toxicology and Environmental Health Part A, 72, 986-997.
56. Mesnage, R., Bernay, B., Séralini, G.-E. (2013). Ethoxylated adjuvants of glyphosate-based herbicides are active principles of human cell toxicity. Toxicology, 313, 122-128.
57. Samsel, A., Seneff, S. (2013). Glyphosate's Suppression of Cytochrome P450 Enzymes and Amino Acid Biosynthesis by the Gut Microbiome: Pathways to Modern Diseases. Entropy, 15(4), 1416-1463.

58. Thongprakaisang, S., Thiantanawat, A., Rangkadilok, N., Tawit Suriyo, T., Satayavivad, J. (2013). Glyphosate induces human breast cancer cells growth via estrogen receptors. Food and Chemical Toxicology, 59, 129-136.

59. Gasnier, C., Dumont, C., Benachour, N., Clair, E., Chagnon, M-C., Séralini, G-E. (2009). Glyphosate-based herbicides are toxic and endocrine disruptors in human cell lines. Toxicology, 262, 184-191.

60. Richard, S., Moslemi, S., Sipahutar, H., Benachour, N., Seralini, G.E. (2005). Differential effects of glyphosate and roundup on human placental cells and aromatase. Environmental health perspectives, 716-720.

61. Koller, V.J., Furhacker, M., Nersesyan, A., Misik, M., Eisenbauer, M., Knasmueller, S. (2012). Cytotoxic and DNA-damaging properties of glyphosate and roundup in human-derived buccal epithelial cells. Archives of Toxicology, 86, 805-813.

5

Genotoxicity of the Neonicotinoid Insecticide Poncho (Clothianidin) on Cd1 Mice Based on Alkaline Comet and Micronucleus Assays

1. Introduction

The neonicotinoid pesticides are nicotine synthetic chemical compounds used in agriculture, veterinary, and human medicine. These pesticides are represented by three groups: nitromethylenes (nithiazine and nitenpyram), nitroguanidines (imidacloprid, thiamethoxam, dinotefuran, and clothianidin), and cyanoamidines (thiacloprid and acetamiprid) [1].

Diverse commercial formulations of neonicotinoid insecticides have been registered in Mexico, such as Jade (imidacloprid), Calypso (thiacloprid), and Poncho (clothianidin). Poncho® is one of the most widely used to treat maize seeds because of its broad-spectrum insecticide activity and relatively small risk to nontarget organisms and the environment. these pesticides have less persistence in the environment due to their low toxicity, no bioaccumulation in biological tissues, and high efficacy to control pests in agricultural crops, they are a replacement for pyrethroids, organophosphates, and carbamate pesticides [2].

Neonicotinoid insecticides act selectively on central nervous system as agonists of the nicotinic acetylcholine receptor (nAChR) in both invertebrates and vertebrates [3]. They are classified to be moderately toxic (category II) and slightly toxic (category III) to mammalian. Reviews of the safety of neonicotinoid insecticides have been performed by several regulatory agencies, and their authors have concluded that there are no indications of any human health effects [4].

Epidemiological studies have suggested a correlation between exposure to pesticides and increase the incidence of diseases and cancer. However, the toxic effects by exposure to pesticides depend on diverse factors, including time and exposure pathways, pesticide type, metabolism, diet, state of antioxidants, and general health [5]. It is important to evaluate the genotoxic actions of these new agricultural pesticides to increase the available toxicological data and to promote regular use that does not pollute the environment and leave pesticide residue in water and food sources where it is a possible risk to human health and the health of other organisms. Exposure to pesticides has also been associated with an increased incidence of functional alterations in the nervous [6], respiratory [7], reproductive [8], and immune systems [9].

Some investigations have demonstrated that clothianidin-based insecticides induce *in vitro* DNA damage in human peripheral lymphocytes [10] and affect the reproduction of rats [11] and birds [12], as well as DNA fragmentation in germinal cells and the inhibition or delay of embryonic development [13]. Because clothianidin represents a health risk and information on its *in vivo* genotoxic effects is scarce, the present study aimed to investigate the DNA damage caused by this agrochemical in the blood peripheral cells of CD1 mice through alkaline comet and micronuclei tests. The comet assay is an early effect biomarker that detects DNA single- and double-strand breaks, or apoptosis between others, induced by environmental genotoxins both *in vitro* [16] and *in vivo* [17].The micronucleus assay is one of the most widely used *in vivo* screening tests in genotoxicity testing [18]. Classically, a micronucleus is defined as a small extranuclear chromatin body originating from a centric fragment or a whole chromosome lost from the metaphase plate. Therefore, frequencies of micronuclei have been considered reliable biomarkers of both chromosome breakage and chromosome loss [18, 19]. Thus, a major strength of the assay is its ability to detect the effects of both clastogens [19] and aneugens [19].

2. Materials and Methods

2.1. Preparation of the neonicotinoid insecticide dilutions

The neonicotinoid insecticide Poncho was donated by Bayer CropScience (México) for the genotoxicity assays. Poncho® (*E*-1-(2-Chloro-1, 3-thiazol-5-ylmethyl)-3-methyl-2-nitroguanidine; RSCO-INAC-103K-301-342-048; flowable suspension: clothianidin, 600 g a.i/L) was diluted with deionized water; cyclophosphamide was also diluted in deionized water. The concentrations of clothianidin used were selected from our preliminary *in vivo* studies in mice using micronuclei and alkaline comet assays.

2.2. Animals

Two- to three-month-old CD1 male mice (~30–35 g body weight) were acclimatized for 2 weeks before the start of the treatments. The mice were housed in polypropylene cages with sawdust bedding at a controlled temperature of 20

± 2°C and a humidity of 50 ± 10% with a 12-h light:12-h dark cycle. They were given a standard diet and water ad libitum in the Faculty of Sciences Biotery at University National Autonomous of Mexico. The animals were treated and housed in accordance with the Guide for the Care and Use of Laboratory Animals.

2.3. Experimental design

Twenty animals were divided into five groups of four animals each and intraperitoneally injected with 20, 40, or 60 mg/kg weight body of Poncho every 3 days for 21 days. A positive control group was intraperitoneally injected with 40 mg/kg weight body of cyclophosphamide, and a negative control group was injected with deionized water under the same conditions. Peripheral blood samples were obtained from the caudal vein of each mouse before treatment and every third day during treatment and used to carry out the alkaline comet and micronucleus assays. Additionally, body weight and food consumption were monitored before and during the treatments.

2.4. Alkaline comet assay

The alkaline comet assay was performed as previously described [14, 15]. Briefly, blood peripheral cells (1 μL) were mixed with 200 μL of low melting point agarose (0.5%) at 37°C, and 100 μL was placed on a frosted slide (Fisher) with normal melting point agarose (1%) and covered with a coverslip. Two gels were made for each animal. The gels were solidified for 5 min at 4°C after they were immersed in a freshly prepared cold lysis solution (2.5 M NaCl, 100 mM EDTA, 10 mM Tris, 1% Triton X-100, and 10% DMSO, pH 10) at 4°C for 1 h. Unwinding the DNA was realized in cold alkaline buffer (300 mM NaOH, 1 mM EDTA, pH 13) for 20 min. Electrophoresis was carried out at 25 V and 300 mA for 20 min. The gels were washed with neutralization buffer (0.4 M Tris, pH 7.5) and fixed with cold absolute methanol for 10 min. The gels were coded and stained with 70 μL of GelRed (10%) and observed in fluorescent microscope (Axiostar Plus Carl Zeiss) with an excitation filter (515–560 nm) and a barrier filter (590 nm) at 40× magnification using the Comet IV software. Three genotoxic parameters were quantified: (a) frequency of comets (nuclei with DNA damage), (b) length of tail (DNA fragmentation), and (c) tail moment in 100 consecutive nuclei (Figure 1)[20].

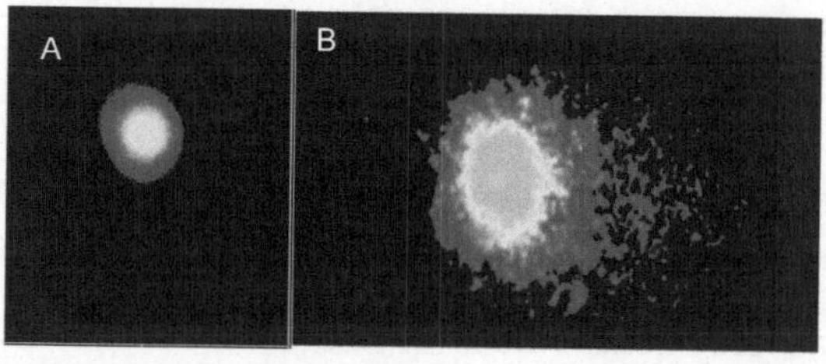

Figure 1. Peripheral blood cell from CD1 male mice. A: Nuclei without DNA damage (without comet) and B: Nuclei with DNA damage (with comet).

2.5. Micronuclei test

Two microliters of peripheral blood was used for blood smears (two for each animal), air-dried, and immediately fixed in absolute methanol and stained with Giemsa (10%) in Sorensen's solution for 6 min. The frequency of micronuclei (MN) was estimated by analyzing 2000 peripheral blood erythrocytes from each animal (1000 per slide) in each group by microscopy at 1000× magnification. The criteria for the identification of micronuclei were based on a standard procedure, and the analysis was restricted to micronuclei in the vicinity of the main nuclei (Figure 2) [18, 19,20].

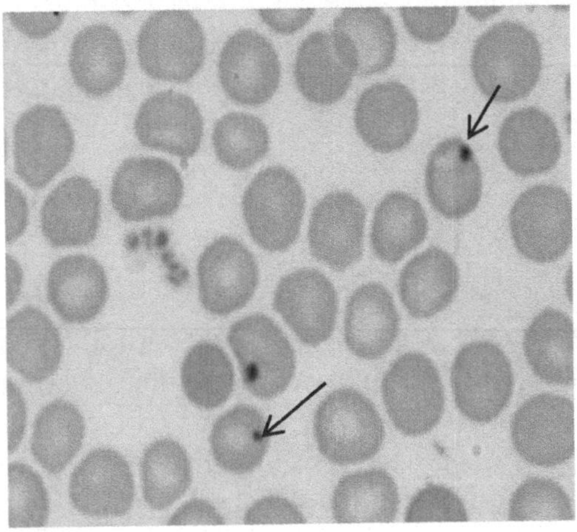

Figure 2. Peripheral blood erythrocytes with micronucleus from CD1 male mice after exposition with Poncho neonicotinoid insecticide.

2.6. Statistical analysis

Comet frequencies, tail lengths, tail moments, and frequencies of micronuclei are reported as the mean ± the standard error of the mean (SEM) obtained for one treatment group. One-way analysis of variance (ANOVA), Student *t* test, and Bonferroni test were used to determine significant differences between the treated groups and the negative and positive control groups using GraphPad Prism. The results were considered statistically significant at $P < 0.05$.

3. Results

3.1. DNA damage induced by Poncho neonicotinoid insecticide in peripheral blood cells of CD1 mice

Table 1 shows the mean values of the comet frequencies, tail lengths, and tail moments measured from the peripheral blood cells of CD1 mice after the intraperitoneal injection of Poncho (40 ,20, and 80 mg/kg body weight every 3 days) over 21 days. The genotoxicity results indicate that after the second administration of 20 mg/kg body weight of Poncho (at 6 days), there was no statistically significant effect on these three genotoxic parameters in the blood peripheral cells compared with either the negative control group at 6 days or basal levels in pretreatment mice. However, after the third administration with 20 mg/kg body weight of Poncho, the mean values of these genotoxic parameters were significantly greater than those of the negative control group and the pretreatment mice. This increase in induced DNA damage in the blood cells remained throughout 21 days of treatment.

After the second administration of 40 mg/kg body weight (at 6 days), the mean values of the three genotoxic parameters were significantly higher in the blood cells of the treated CD1 mice than in the negative control group or in the pretreatment mice. Furthermore, after the third intraperitoneal administration, the mean comet frequency, tail length, and tail moment were greatly increased compared with the negative control group and the pretreatment mice. A higher frequency of comets with greater tails occurred after sixth administrations (at 15 days), and apoptotic nuclei were observed after the seventh administration (results not shown).

In CD1 mice treated with 80 mg/kg body weight of Poncho insecticide, the means of the three genotoxic parameters significantly increased after the first administration (at 3 days) and remained higher than those of the negative control group and the pretreatment mice throughout the treatment (Table 1). This high concentration of insecticide produced severe DNA damage in the blood peripheral cells after three treatments, with 70% of cells producing comets and larger tail lengths and tail moments than the negative control group or the pretreatment mice. However, the mean comet frequency, tail length, and tail moment were significantly decreased after 15 days of administration compared with the negative control group and the pretreatment mice. Greater numbers of apoptotic nuclei were observed throughout the course of treatment (data not shown).

The micronuclei frequencies in the peripheral blood erythrocytes of CD1 male mice treated with the Poncho insecticide are shown in Table 2. No significant increase in micronuclei frequency was observed in the blood cells of the male mice after the first administration with 20 mg/kg body weight Poncho compared with either negative control or pretreatment mice. However, after the second administration, the frequency of micronuclei was significantly higher in the treated mice than in the negative controls or the pretreatment mice ($P < 0.05$). This significant induction of micronuclei was observed and remained constant throughout the treatment course.

Table 1. DNA damage induced by Poncho neonicotinoid insecticide in peripheral blood cells of CD1 male mice

	Animals	**Genotoxic**	**Repeted exposition**		**(after days)**					
Concentration	**(n)**	**parameters**	day 0	3	6	9	12	15	18	21
of Poncho		means ± SEM								
mg/Kg										
		frequency of								
20	4	comets	4.0 ± 1.9	6.0± 0.8	4.0± 1.0	12.0 ±1.8*	16.0 ± 2.1*	24.0 ± 1.6**	36.0 ± 2.4**	44.0 ± 1.0**
		Tail lenght	16.33 ± 0.9	16.77± 0.9	18.22 ± 0.9	34.00 ± 0.9**	36.44± 2.4**	40.67± 2.9**	44.31± 2.2**	54.22 ± 2.2**
		Tail moment	0.36 ± 1.2	0.39 ± 0.4	0.56± 0.6	2.11 ± 0.9**	2.44± 2.9**	2.99 ± 2.0**	3.33± 2.0**	3.88 ± 2.0**
		frequency of								
40	4	comets	4.0 ± 1.1	6.0± 1.2	18.00±1.0*	34.0 ± 0.6*	46.0 ± 2.0*	52.0 ± 2.2**	42.0 ± 1.3**	40.0 ± 1.9**
		Tail lenght	18.44 ± 0.9	21.08± 1.3	36.22±1.3**	48.00 ± 2.9**	36.44± 0.9**	60.67± 2.4**	44.31± 2.2**	54.22 ± 2.2**
		Tail moment	0.36 ± 0.9	0.88 ± 1.1	2.96 ± 1.9**	2.61 ± 1.4**	2.44± 1.2**	3.69 ± 1.6**	3.33± 1.0**	3.28 ± 1.2**
80	4	frequency of								
		comets	4.0 ± 1.9	16.0 ± 0.8*	36.0± 1.0**	50.0 ± 0.6**	71.0± 1.4**	64.0± 1.2**	50.0 ± 0.9**	54.0 ± 2.1**
		Tail lenght	12.22 ± 0.9	36.77 ± 2.1*	43.22 ± 0.9**	66.90 ± 1.9**	86.48 ± 2.2**	62.90± 2.4**	64.66± 2.4**	52.82 ± 1.8**

	Animals	Genotoxic	Repeted exposition		(after days)					
		Tail moment	0.64 ± 1.4	3.12 ± 2.0*	3.99 ± 0.9**	5.66 ± 1.5**	6.88 ± 2.7**	4.99 ± 1.9**	5.12± 2.9**	4.67 ± 2.9**
Negative	4	frequency of								
Control		comets	6.0± 1.9	2.0± 0.8	4.0± 1.0	6.0± 0.6	6.0 ± 07	8.0 ± 1.0	6.0 ± 1.0	4.0± 2.0
Group		Tail lenght	14.33 ± 0.8	22.77± 2.4	23.11 ± 2.4	18.40 ± 2.1	16.24± 0.9	20.99± 1.4	14.31± 0.8	22.33 ± 1.4
		Tail moment	0.52 ± 1.2	0.77 ± 0.8	0.83 ± 1.0	0.99± 1.2	1.44± 0.8	0.99 ± 0.5	0.86± 0.9	0.78 ± 1.6
CP		frequency of								
Positive	4	comets	4.0 ± 1.2	10.0 ± 0.8*	18.0 ± 1.0**	24.0 ± 0.6**	30.0 ± 0.9**	56.0 ± 1.4**	44.0 ± 2.7**	34.0 ± 2.3**
control		Tail lenght	18.48 ± 1.6	26.77± 2.2*	33.22 ± 2.4**	38.00 ± 1.2**	42.44± 2.2**	70.67± 2.3**	44.31± 0.9**	34.82 ± 1.8**
		Tail moment	0.72 ± 1.2	1.99 ± 1.9*	3.66 ± 2.3**	3.88 ± 2.1**	5.44± 2.0**	4.56 ± 2.0**	3.99± 1.2**	3.24 ± 0.4**

Genotoxic parameters means =100 nuclei for organism in each group

*Significant differences among treated groups by analysis of Student ´t° test $p < 0.05$, and therefore the Bonferroni comparison test was applied, $p < 0.01$ **

CP=40 mg/kg of cyclophosphamide

3.2. Micronuclei frequency in peripheral blood erythrocytes from CD1 mice treated with Poncho insecticide

Table 2. Micronuclei frequency in perypheral blood erythrocytes from CD1 male mice treated with Poncho neonicotinoid insecticide

	Animals	Treatment squeme		(after days)					
Concentration	(n)	day 0	3	6	9	12	15	18	21
of Poncho		Frequency of micronuclei (means ± SEM)							
mg/Kg									
20	4	0.5 ± 0.91	0.7 8± 0.82	1.66 ± 0.70*	2.94 ± 0.82*	2.67 ± 0.39*	2.44 ± 0.66*	3.69 ± 1.04*	5.46 ± 1.04*
40	4	0.8 ± 0.86	0.92 ± 0.23	2.33 ± 0.42*	3.82 ± 0.44*	4.99 ± 0.99*	8.66 ± 0.79*	14.0 ± 1.09*	16.0 ± 1.02*
80	4	0.7 ± 0.33	0.74 ± 0.84	3.55 ± 0.48*	5.45 ± 0.62*	7.0 ± 0.49*	12.0 ± 0.99.*	18.0 ± 0.99*	24.0 ± 0.93*
Negative									
Control	4	0.8 ± 0.93	0.92 ± 0.80	0.57 ± 1.09	1.0± 0.69	0.78 ± 0.79	1.0 ± 0.88	1.0 ± 0.90	1.0± 0.76
Group									
positive									
Control	4	0.5 ± 0.29	4.96 ± 0.83*	5.88 ± 1.00*	7.99 ± 1.01*	9.78 ± 1.10*	13.09± 1.42*	13.0 ± 1.22*	14.0 ± 1.02*
Group									
CP									

n=2000 peripheral blood erythocytes forn organism in each group

*Significant differences among treated groups by analysis of Student °t° test were found the $p < 0.05$, and therefore the Bonferroni comparison test was applied, $p < 0.01$

CP= 40 mg/Kg of cyclophosphamide

At concentrations of 40 and 80 mg/kg body weight, no significant differences in micronuclei frequency compared with controls or pretreatment mice were observed after the first administration first ($P < 0.05$). However, after the second administration, both 40 and 80 mg/kg body weight of Poncho significantly increased the micronuclei frequency in treated mice compared with negative control or pretreatment mice ($P < 0.05$). The observed production of chromosomal damage was consistent in all of the groups after administration throughout the treatment course (Table 2).

Cyclophosphamide, which was used as a positive control, produced significant increases in DNA strand breaks and the frequency of micronuclei in the blood peripheral cells of treated CD1 male mice compared with the negative control and pretreatment mice. An increase in the percentage of micronuclei was observed after the second injection and throughout the treatment course (Tables 1 and 2).

4. Discussion

The present study was designed to evaluate DNA damage using alkaline comet assay and measurements of micronuclei frequency in CD1 male mice treated every 3 days over a 21-day period with low (20 mg/kg body weight) and high concentrations (40 and 80 mg/kg body weight) of the neonicotinoid insecticide Poncho (clothianidin). We found that clothianidin induced concentration- and time-dependent increases in DNA strand breaks and micronuclei frequency in the peripheral blood cells of CD1 mice. Our results are in agreement with data obtained from *in vivo* genotoxic studies performed with other neonicotinoid insecticides. In Wistar albino rats intraperitoneally injected with 50 or 100 mg/kg body weight of commercial acetamiprid and imidacloprid formulation, Confidor® and pure imidacloprid for 90 days induced a significant increase of frequency of micronuclei and chromosomal aberrations in the bone marrow cells [21]. Therefore, this increase in micronuclei frequency was observed at 300 mg/kg body weight of imidacloprid (Confidor®) in rat bone marrow cells [22]. At 0.2 or 0.5 mg/kg imidacloprid in dry soil for 14 days produced spermatic malformations in *Eisenia fetida*. The same species of nematode exposed to 0.05, 0.1, 0.2, or 0.5 mg/L imidacloprid for 2 h caused a significant increase in DNA damage in coelomocytes with relationship response doses [23, 24]. Also, an increase in the frequency of micronuclei and DNA strand breaks were observed in peripheral blood erythrocytes of *Rana N. Hallowell* exposed to 8 ,0.5 ,0.2 ,0.1 ,0.05, or 32 mg L^{-1} imidacloprid (pure compound) for 7 days [25].

When comparing the genotoxic action of the three tested concentrations of Poncho insecticide in CD1 male mice, we observed that at low concentrations (20 mg/kg body weight), DNA damage occurred after the third administration, and the genotoxic effect was constant throughout the treatment course. This increase was also observed at concentrations of 40 and 80 mg/kg body weight of insecticide, although these concentrations produced even higher DNA damage at a few time points during the treatment course. Additionally, we observed that the long-term administration of high concentrations (40 and 80 mg/kg body weight) of clothianidin produced an apoptotic response in the peripheral blood cells of this animal model. Perhaps this apoptotic response was associated with the increased induction of DNA damage in the blood cells, as well as a mechanism to eliminate damaged cells from the blood tissue. Another interesting observation in this study was that the increased DNA damage observed upon treatment with either concentration of Poncho insecticide injected every third day persisted for at least 48 h. These results indicate that the observed DNA damage was not repaired and was possibly caused by excessive reactive oxygen species (ROS) production or greater accumulation of the pesticide or intermediate metabolites that interfered with the repair mechanisms of the mice or interacted with DNA to produce the fragmentation visualized through micronuclei and comet formation.

The genotoxic and apoptotic actions induced *in vivo* by Clothianidin in peripheral blood cells may be related to the following factors: (a) the penetration or absorption of the pesticide in the body and the metabolic pathways that differentially produce intermediate metabolites, (b) the generation and overall levels of ROS or free radicals that subsequently inactivate detoxification mechanisms leading to increased oxidative stress, and (c) alterations in DNA damage repair mechanisms. Another probable explanation for the effects is the accumulation of the original molecule, which may produce DNA single-strand breaks and induce the formation of micronuclei in the blood tissue.

The biotransformation of clothianidin insecticide has been studied in rats [26] and mice [27, 28]. The *in vivo* metabolic pathways of clothianidin in the liver microsomes of rats have been described. The major metabolic reactions (Phase I metabolism) are oxidative demethylation by microsomal CYP450-enzymes to form *N*-(2-chlorothiazol-5-ylmethyl)-*N*-nitroguanidine [28] During Phase II metabolism, it is metabolized mainly to *N*-methyl-*N*-nitroguanidine and 2-(methylthio) thiazole-5-carboxylic acid via glutathione conjugation and other guanidine derivatives [27, 28]. Perhaps, some of these metabolic products are involved in the genotoxicity of clothianidin in CD1 male mice.

In general, the involvement of ROS has been postulated as a possible mechanism of the genotoxicity and oxidative damage to cellular molecules such as lipids, proteins, and nucleic acids caused by neonicotinoid pesticides [29]. Rats treated with 32 mg/kg body weight of clothianidin by gavage for 90 days exhibited significantly increased sperm DNA fragmentation and apoptosis caused by oxidative stress [11, 12]. Similar effects were observed in the germinal cells of male and female quails exposed to 50 mg/kg body weight of clothianidin [13].

Poncho neonicotinoid insecticide is a genotoxic agent that is likely to be a source of free radicals or ROS in CD1 mice. ROSs, such as superoxide anions ($O_2\bullet$), hydrogen peroxide (H_2O_2), and hydroxyl radicals (OH–), are highly reactive with DNA and produce DNA damage, including single- and double-strand breaks and micronuclei formation. The increased DNA damage detected in the blood peripheral cells of the CD1 mice in this study could be considered potential premutagenic lesions, which may increase the risk of cancer and contribute to autoimmune diseases and other alterations [30].

We used cyclophosphamide as a positive control because it is a potential genotoxic agent in experimental animals [31]. It is a DNA alkylating agent and a good inducer of oxidative stress. Cyclophosphamide is metabolized to 4-hydroxycyclophosphamide, aldophosphamide mustard, and acrolein by cytochrome P-450 enzymes, which are involved in the increase of oxidative stress, DNA damage, and apoptosis in mammalian cells and produce cytotoxicity in various target organs [32].

In our research, cyclophosphamide induced a significant increase in the three genotoxic parameters measured as well as an increase in the frequency of micronuclei in the peripheral blood cells of CD1 male mice. These results are in agreement with and reported in numerous *in vivo* genotoxic studies of cyclophosphamide, which induced sister chromatid exchange, micronucleus, DNA strand breaks, and other genetic alterations [33].

The DNA damage observed in the blood tissue of CD1 male mice treated with both low and high concentrations of neonicotinoid insecticide Poncho suggests potential health risks to humans and animals. Furthermore, the data obtained with two genetic assays corroborate the *in vivo* genotoxic action of this neonicotinoid insecticide and support its regular use since this insecticide does not pollute the environment or leave pesticide residue in water and food sources where it is a possible risk to ecosystems.

References

1. Tomizawa M, Casida JE. Neonicotinoid insecticide toxicology: mechanism of selective action. Annual Review of Pharmacology and Toxicology. 2005; 45: 247–268.
2. Bayer CropScience México. 2008.
3. Millar NS, Denholm I. Nicotinic acetylcholine receptors: targets for commercially important insecticides. Invertebrate Neurosciences. 2007; 1:53–66.
4. Environmental Protection Agency, USA. Clothianidin; Pesticide Tolerance. 2003.
5. Bolognesi C. Genotoxicity of pesticides: Review of human biomonitoring studies. Mutation Research/Genetic Toxicology and Environmental Mutagenesis. 2003; 543: 251–271.
6. Dick FD. Parkinson's diseases and pesticides. British Medical Bulletin. 2007; 80: 219–231.
7. Hoppin JA, Umbach DM, London SJ, Henneberger PK, Kullman GJ. Pesticides and atopic and nonatopic asthma among farm women in the agricultural health study. American Journal of Respiration and Critical Care Medicine. 2008; 177:11–18.
8. Sharpe RM. How strong is the evidence of a link between environmental chemicals and adverse effects on human reproductive health? British Medical Journal,2004; 328: 447–451.
9. Colosio C, Birindelli S, Corsini E, Galli CL, Maroni M. Low level exposure to chemicals and immune system. Toxicology and Applied Pharmacology. 2005; 207: S320–S328.
10. Calderón-Segura ME, Gómez-Arroyo S, Villalobos-Pietrini R, Martínez-Valenzuela C, Carbajal-Carbajal-López Y, Calderón-Ezquerro MC, Cortés-Eslava J, García-Martínez R, Flores-Ramírez D, Rodríguez-Romero MI, Méndez-Pérez P, Bañuelos-Ruíz E. Evaluation of genotoxic and cytotoxic effects in human peripheral blood lymphocytes exposed in vitro to neonicotinoid insecticides news. Journal of Toxicology. 2012; 1: 11. DOI:10.1155/2012/612647

11. Bal R, Türk G, Yılmaz O, Etem E, Kuloğlu T, Baydaş G, Naziroğlu M. Effects of clothianidin exposure on sperm quality, testicular apoptosis and fatty acid composition in developing male rats. Cell Biology and Toxicology. 2012; 28: 187–200. DOI 10.1007/s10565-012-9215-0.

12. Hoshi N, HiranoT, Omotehara T, Tokumoto J, Umemura Y, Mantani Y, Tanida T, Warita K, Tabuchi Y, Yokoyama T, Kitagawa H. Insight into the mechanism of reproductive dysfunction caused by neonicotinoid pesticides. Biological and Pharmaceutical Bulletin. 2014; 37: 1439–1443.

13. Tokumoto J, Danjo M, Kobayashi Y, Kinoshita K, Omotehara T, Tatsumi A, Hashiguchi M, Sekijima T, Kamisoyama H, Yokoyama T, Kitagawa H, Hoshi N. Effects of exposure to clothianidin on the reproductive system of male quails toxicology. Journal of Veterinary Medical Science. 2003; 75: 755–760. DOI: 10.1292/jvms.12-0544.

14. Singh NP, McCoy MT, Tice RR, Schneider EL. A simple technique for quantitation of low levels of DNA damage in individual cells. Experimental Cell Research. 1998; 175: 184–191.

15. Tice RR, Agurell E, Anderson D, Burlinson B, Hartmann A. Single cell gel/comet assay: guidelines for in vitro and in vivo genetic toxicology testing. Environmental Molecular and Mutagenesis. 2000; 35: 206–221.

16. Calderón-Segura ME, López L, Zúñiga R, Sánchez JM, Gómez-Arroyo S. Metabolic activation of herbicides products by Vicia faba detected in human peripheral lymphocytes using alkaline single cell gel electrophoresis. Toxicology In Vitro. 2007; 19: 243–251.

17. Grover P, Danadevi K, Mahboob M, Rozati R, Banu BS, Rahman MF. Evaluation of genetic damage in workers employed in pesticide production utilizing the comet assay. Mutagenesis. 2003; 18: 201–205.

18. Hayashi M, Sofuni T, Ishidate M. High-sensitivity in micronucleus induction of a mouse strain (MS). Mutation Research. 1982; 105: 253–256.

19. Hayashi M, Macgregor JT, Gatehouse DG, Adler ID, Blakey DH, Dertinger SD, Krishna G, Morita T, Russo A, Sutou S. In vivo rodent erythrocyte micronucleus assay. II. Some aspects of protocol design including repeated treatments, integration with toxicity testing, and automated scoring. Environmental and Molecular Mutagenesis 2000; 35:234–252.

20. Hayashi M, Kodama Y, Awogi T, Suzuki T, Asita AO, Sofuni T. The micronucleus assay using peripheral blood reticulocytes from mitomicin C- and cyclophosphamide-treated rats. Mutation Research. 1992; 278: 209–213.

21. Costa C, Silvari, V, Melchinib A, Cataniab S, Heffronc JJ. Genotoxicity of imidacloprid in relation to metabolic activation and composition of the commercial product. Mutation Research. 2009; 672: 40–44.

22. Demsia G, Vlastos D, Goumenou M, Matthopoulos DP. Assessment of the genotoxicity of imidacloprid and metalaxyl in cultured human lymphocytes and rat bone-marrow. Mutation Research. 2007; 634: 32–39.

23. Gomez-Eyes JL, Svendsen C, Lister L, Martin H, Hodson ME, Spurgeon DJ. Measuring and modelling mixture toxicity of imidacloprid and thiacloprid on Caenorhabditis elegans and Eisenia fetida. Ecotoxicology and Environmental Safety. 2009; 72: 71–79.
24. Zang Y, Zhong Y, Luo Y, Kong ZM. Genotoxicity of two novel pesticides for the earthworm, Eisenia fetida. Environmental Pollution. 2000; 108: 271–278.
25. Feng S, Kong Z, Wang X, Zhao L, Peng P. Acute toxicity and genotoxicity of two novel pesticides on amphibian, Rana N. Hallowell. Chemosphere. 2004; 56: 457–463.
26. [26]Yokota T, Mikata K, Nagasaki H, Ohta K. Absorption, tissue distribution, excretion, and metabolism of clothianidin in rats. Journal of Agricultural and Food Chemistry. 2003; 51: 7066–7072.
27. Ford KA, Casida JE. Chloropyridinyl neonicotinoid insecticides diverse molecular substituents contribute to facile metabolism in mice. Chemical Research in Toxicology, 2006; 19: 944–951.
28. Casida JE. Neonicotinoid metabolism compounds, substituents, pathways, enzymes, organisms, and relevance. Journal of Agricultural and Food Chemistry. 2011; 59: 2923–2931.
29. Zhang JJ, WangY, Xiang HY, Li M, Li X, Ma WH, Wang KG, Zhang JH. Oxidative stress: role in acetamiprid-induced impairment of the male mice reproductive system. Agricultural Sciences in China. 2011; 10: 786–796.
30. Valko M, Rhodes CJ, Moncol J, Izakovic M, Mazu M. Free radicals, metals and antioxidants in oxidative stress-induced cancer. Chemico-Biological Interactions. 2006; 160: 1–40.
31. Roy P, Yu LJ, Crespi CL, Waxman DJ. Development of a substrate-activity based approach to identify the major human liver P-450 catalysts of cyclophosphamide and ifosfamide activation based on cDNA-expressed activities and liver microsomal P-450 profiles. Drug Metabolism and Disposition. 1999; 27:655–666.
32. Korkmaz A, Topal T, Oter S. Pathophysiological aspects of cyclophosphamide and ifosfamide induced hemorrhagic cystitis; implication of reactive oxygen and nitrogen species as well as PARP activation. Cell Biology and Toxicology. 2007; 12: 27–34.
33. Anderson D, Bishop JB, Garner RC, Ostrosky-Wegman P, Selby PB. Cyclophosphamide: review of this mutagenicity for an assessment of potential germ cell risk. Mutation Research. 1995; 330: 115–181.

6

The Ecotoxicity of Pyrimethanil for Aquatic Biota

1. Introduction

Increasing food requirements exert a constant pressure for intensifying agricultural activities, recognized, nowadays, as one of the most important economic activities in many high and low income countries [1]. In fact, agriculture has been considered a feasible solution for reducing the levels of poverty and hunger given that the vast majority of poor people in developing countries are concentrated in rural areas [2]. The high demand for agricultural products requires optimizing the production to reduce the loss due to crop diseases such as those caused by fungi. Although incentives for agriculture optimization and development are usually paralleled by sustainable practices, intensive agricultural practices and the pursuit for more profitable productions have unfortunately escalated the increase in the use of agrochemicals against crop pests/pathogens. Additionally, the agrochemical market represents an important economic sector for many countries [3]. According to the previously mentioned authors, although the use of chemicals such as lime sulfur and Bordeaux mixture as fungicides began in the mid-1800s, only in the 1960s did fungicides with specific (systemic) modes of action become protagonists in controlling against fungal pathogens. The more serious consequence is the fact that the impact of agrochemicals is not only on pests and pathogens, but also on non-target organisms inhabiting adjacent areas, including humans. The excessive and indiscriminate application of agrochemicals linked to a lack of legal control about their use, commercialization and regularization in many countries has given agrochemicals a primary role of concern in environmental management. Among the groups of chemicals used in agriculture against pathogens, fungicides are the third most used agrochemical group, representing ca. 23% of sales on the agrochemical market [4]. Contrary to most agrochemicals, fungicides are frequently applied in a prophylactic manner several times per season, although at lower application rates than most herbicides and insecticides, which increases the risk of chronic exposure to aquatic biota [5]. On the other hand, some organisms can develop resistance to fungicides after relatively short periods (years) of exposure, resulting in fungal pathogens being responsible for important economic losses of fruit and vegetable products [5, 6].

Via the application of agrochemicals, farmers currently guarantee high productivity of fruit and vegetable crops. However, reduction of crop losses by using excessive amounts of such chemicals has a negative effect on aquatic organisms. The spray-drift, leaching, run-off or accidental spills occurring during or after application of agrochemicals has become a serious and increasing worldwide problem for aquatic ecosystems [7, 8]. Pyrimethanil (PYR) is one of the most used fungicides that has been detected in many aquatic ecosystems [5] and one of the most frequently used in European vineyards [9, 10, 11]. Such increase has heightened the interest in studying the potential risk and influence of PYR on the environment [3, 5, 12–14].

The main objective of this chapter is to provide information on the environmental risks posed by PYR for aquatic organisms. For this, PYR chemical characteristics as well as its potential risk for the aquatic environment will firstly be provided and subsequently three different approaches will be discussed: (i) assessment of toxic effects of the pure active ingredient or the commercial formulation on primary producers using traditional assays with forced exposure, (ii) assessment of toxic effects of the pure active ingredient and PYR formulation on aquatic animals using traditional assays with forced exposure and *in situ* experiments, and (iii) estimation of the role of PYR as an environmental disturber by triggering avoidance response in a non-forced exposure system.

2. Pyrimethanil: Characteristics and Hazard Potential

The fungicide PYR (*N*-(4, 6-dimethylpyrimidin-2-yl)-aniline; CAS number 53112-28-0) is an anilinopyrimidine fungicide that inhibits the secretion of fungal enzymes produced in the infection process [15, 16]. It was recently developed to act on resistant fungi strains, mainly to control *Botrytis cinerea* in grapes (wine), *Venturia inaequalis* in apples and *Botrytis* spp. in protein peas [17], reason for which its use has increased greatly [18, 19]. PYR rapidly penetrates the cuticle and inhibits the secretion of fungal enzymes required for the infection process, blocking the ability of fungi to degrade and digest the plant tissues, and thus stopping development of the disease [17, FAO, see http://www.fao.org/ fileadmin/ templates/agphome/documents/Pests_Pesticides /JMPR/ Evaluation07/ Pyrimethanil.pdf]. The commercial products that contain PYR as active ingredient are Clarinet®, Mythos®, Rubin®, Scala®, Siganex®, Vision®, and Walabi®, which are currently used both pre- and post-harvest to protect various crops such as apple, banana, carrot, citrus, grape, melon, onion, potato, strawberry, and tomato [16, 18, 20, see also www.bayercropscience.com.br/Site/nossosprodutos/

protecaodecultivosebiotecnologia/DetalheDoProduto.fss?Produto=44]. According to the EFSA report [17], PYR is rapidly excreted once orally absorbed, it has no potential to bioaccumulate, has a low acute toxicity, is not teratogenic and seems to have no neurotoxic effect; however, studies have observed acute and chronic toxicity for non-target organisms [7, 14, 21–27] that converts PYR into an environmental disturber of concern. Unfortunately, despite the intensive agricultural use of PYR, there is not an exhaustive study regarding the effects on adjacent aquatic ecosystems. This is possibly related to the assumption that PYR has a short half-life, with fast degradation [17, 28] and, therefore, possible toxic effects may occur at short term but are minimized at mid- and long-term. Some chemical and (eco)toxicological characteristics of PYR published by EFSA [17] can be seen in Table 1.

The regulatory decision to approve a given agrochemical should be based not only on its efficiency in controlling the pest/pathogen, but also on the potential environmental impact on non-target organisms inhabiting both target and nearby areas. Given the lack of information on PYR biological effects and the imminent need to expand the range of toxicity tests, a series of recent studies have been conducted to fill in this information gap. This chapter is to present these results and integrate them with the information that was already available. With the latter purpose, different aquatic organisms from different levels of biological organization have been used in ecotoxicological studies in the past years to evaluate the potential risks due to exposure to PYR via different routes. Beside traditional toxicity tests, approaches taking into account multigeneration responses, temperature influence and behavioral endpoints as well as *in situ* exposures have been performed and will be discussed in the next sections.

Table 1. Chemical and (eco)toxicological characteristics of pyrimethanil [17].

Chemical name (IUPAC)	*N*-(4, 6-dimethylpyrimidin-2-yl) aniline
Chemical name (CA)	4, 6-dimethyl-*N*-phenyl-2-pyrimidinamine
Molecular formula	C12H13N3
Molecular mass	199.28 g mol-1
Structural formula	

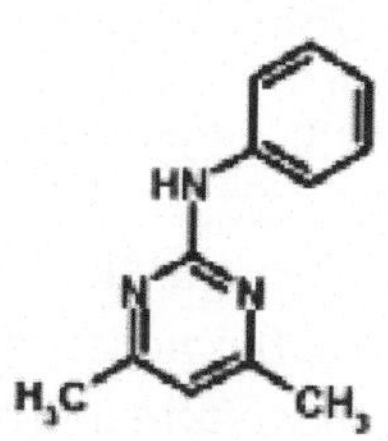

Temperature of decomposition		189.54 to 344.74 °C
Flammability		Not flammable

Explosive properties		Not explosive
Skin irritation		Not irritating
Eye irritation		Not irritating
Genotoxicity		No evidence
Degradation time in water	DT50 water	8.9 to 24 days
and sediment	DT90 water	70 to 99 days
	DT50 whole system	40 to 121 days
	DT90 whole system	Not stated and 134 days
Toxicity for aquatic	Rainbow trout	LC50 (96 h): 10.56 mg L-1
Organisms	*Daphnia* sp.	EC50 (96 h): 2.9 mg L-1
	Green alga*	EbC50/ ErC50 (96 h): 1.2/5.84 mg L-1
	Daphnia magna	NOEC (reproduction, 21 d): 0.94 mg L-1
	Chironomus riparius	NOEC (emergence, 28 d): 4.0 mg L-1
Ecotoxicological data		Harmful

- CA: Chemical Abstract; DT_{50} and DT_{90}: period required for 50% and 90% dissipation; EC_{50}: median effective concentration; EbC_{50}: the concentration at which 50% reduction of biomass is observed, ErC_{50}: the concentration at which 50% reduction of growth rate is observed, IUPAC: International Union of Pure and Applied Chemistry; LC_{50}: median lethal concentration; NOEC: no observed effect concentration. * Species name is not provided.

3. Toxicity of Pyrimethanil to Primary Producers (Microalgae and Macrophytes)

Agrochemicals can considerably affect the structure of algal communities generating functional changes, due to alterations in biotic interactions. Freshwater macrophytes and microalgae usually are not the target of agrochemicals; however, the potential impact that these compounds can have on primary producers is well known [29]. Various studies alerting to the risk of excessive excessive pesticide application and consequent pollution of aquatic environments have been performed using different microalgae, duckweeds, and the aquatic plant *Myriophyllum aquaticum* as test organisms [8, 30–36].

The growth of the floating plants *Lemna minor* and *L. gibba* was inhibited by pure PYR with an IrC_{50} of 23 and 7.8 mg L^{-1}, respectively [23, 28]. In the same concentration range, the growth of the unicellular green algae *Scenedesmus acutus, S. obliquus, Desmodesmus subspicatus,* and *Raphidocelis subcapitata* (=*Pseudokirchneriella subcapitata* and also *Selenastrum capricornutum*) was affected [7, 23, 25, 28]. On the other hand, the diatom species *Gomphonema parviculum* revealed lower PYR effect concentrations (IrC_{50} = 0.24 mg L^{-1}, unpublished data). After 4-days-exposure to 0.2 mg L^{-1} PYR, maximal photosynthetic capacity of the macrophytes *Callitriche palustris* and *Elodea canadensis* was significantly inhibited [21].

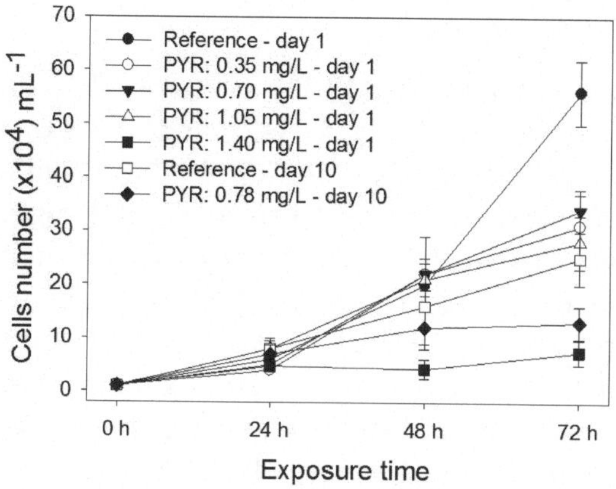

Figure 1. Growth curves of the microalgae P. subcapitata after 72 h exposure to reference and pyrimethanil-treated mesocosm samples taken at days 1 and 10 post-application.

Also, toxicity tests with water from PYR-treated mesocosms (commercial formulation Mythos®; initial PYR concentration of 1.4 mg L^{-1}) were performed with the microalgae *P. subcapitata* by Shinn et al. [14]. In the latter study, water samples were taken at day 1 and 10 after PYR application to verify whether immediate effects and those after subsequent PYR dissipation, respectively, would be different. The cell growth during 72 h exposure to 1-d PYR-treated mesocosm water diluted at different concentrations and to the 10-d undiluted PYR-treated mesocosm sample is summarized in Figure 1, together with the growth of cells exposed to water from the reference mesocosm. *P. subcapitata* cells exposed to samples from PYR-treated mesocosms taken at day 1 had their growth strongly inhibited, whereas the effects on cell growth were less pronounced at day 10, when PYR concentrations showed a reduction of 0.78) %45 mg L^{-1}) in relation to day 1 (1.40 mg L^{-1}). According to Shinn et al. [14], these results indicate that depending on the concentration at which PYR reaches the affected aquatic system, immediate effects on phytoplankton species can be observed. However, assuming an interruption of the contamination by the fungicide, these same effects may become considerably attenuated in view of the dissipation of PYR from the aquatic compartment.

4. Toxicity of Pyrimethanil to Aquatic Animals

The toxic effects of pure PYR on different aquatic animals and responses have also been addressed. For the cladocera *Daphnia magna* the 96 h LC_{50} (lethal concentration to %50 of exposed organisms) of pure PYR ranged from 1.2 to 2.9

mg L^{-1} and the no observed effect concentration (NOEC) on reproduction after 21 d exposure from 0.5 to 0.9 mg L^{-1} [17, 23]. Surprisingly, *D. magna* exposed to 1.0 mg L^{-1} PYR (LOEC for reproduction, 21 d) at 16 ,14, and 19 °C did not produce a F_1-generation [23]. The EC_{50} for the reproduction of the sister species *D. pulex* was 0.69 mg L^{-1} and the NOEC was 0.015 mg L^{-1} [25].

Regarding aquatic insects, the NOEC of pure PYR for the non-biting midge *Chironomus riparius* was 4 mg L^{-1} [22] and the EC_{50} for the phantom midge *Chaoborus flavicans* was 1.78 mg L^{-1} [25]. Within a similar PYR range, the oligochaete *Lumbriculus variegatus* revealed a NOEC of 4 mg L^{-1} with respect to reproduction [23] and the snail *Physella acuta* presented embryo LC_{50} of 0.402 mg L^{-1} [24]. The lethal PYR concentration for 50% of a rainbow trout population (*Oncorhynchus mykiss*) was 14 mg L^{-1}, whereas the NOEC for the parameter dry weight was 0.07 mg L^{-1} PYR [37]. Survival and oxidative stress of the aquatic worm *Tubifex tubifex* has also been used as an endpoint to assess the toxicity of PYR [12]. Compared with results previously described, LC_{50} values were relatively high, between 39 and 49 mg L^{-1}, after 7 and 1 day exposures respectively. On the other hand, effects on the activity of catalase (increased activity) and glutathione-*S*-transferase (decreased activity) were observed at lower (25 mg L^{-1}) concentrations [12]. These same authors detected a quick (after 4 d) bioaccumulation of PYR in the worm, but that was reduced in the subsequent days.

Recently, impairment in the feeding ability of the tropical cladoceran *Ceriodaphnia silvestrii* was verified by Araújo and collaborators (unpublished data) using the post-exposure feeding endpoint, increasingly used in ecotoxicological studies [38–41]. In the study of Araujo and collaborators (unpublished data), a PYR contamination scenario was simulated by applying the commercial formulation Mythos® to a mesocosm system. Two treatments were considered: reference (non-contaminated) mesocosms and mesocosms contaminated with PYR at 1.40 mg L^{-1}. The *in situ* exposure started one day after application, when *C. silvestri* individuals (3rd brood; 3 days old) were exposed for 24 h in the mesocosms in 250 mL cylindrical chambers as shown in Figure 2. After this period organisms were removed, transported to the laboratory in containers with the respective mesocosm water, checked for mortality, and surviving *C. silvestri*were fed with a *P. subcapitata* algal suspension of known concentration for 4 h, time after which remaining cells were recorded.The feeding of individuals exposed to the commercial formulation Mythos® applied to mesocosm systems was inhibited up to %12 ± 31, with no lethal effect. The absence of an acute effect was expected as according to published data the lethal toxicity of PYR for the crustacean *D. magna* is known to occur at concentrations around 3 mg L^{-1} [37] and chronic toxicity for *D. magna* reproduction at approximately 1 mg L^{-1} [23]. The PYR effects on the feeding of *C. silvestri* showed post-exposure feeding can be a suitable endpoint to discriminate the effects of contamination caused by the fungicide PYR. Since feeding is a mechanism by which organisms obtain energy for many biological functions (e.g., development, growth, reproduction, survival), the impairment of the feeding capacity may be critical not only for the organisms themselves (reduced competitive ability, energy imbalance, higher susceptibility to predation), but also for ecosystem functioning as an unbalance in the food web (increase in phytoplankton community due to a decrease in grazing) may have consequences at the community level [39].

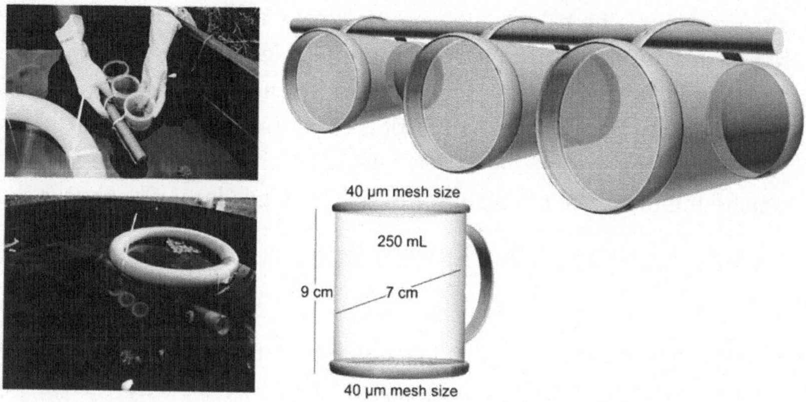

Figure 2. Exposure of Ceriodaphnia silvestri to pyrimethanil in mesocosms and schematic diagram of the exposure chamber.

Aiming to assess the suitability of the yeast *Saccharomyces cerevisiae* gene expression–based assay for screening the toxicity of worst-cases of soil and water PYR contamination, soils were sprayed with the commercial formulation Scala® to simulate accidental spill doses and runoff events [42]. The authors found that, although less sensitive, the yeast-based assay correlated well with the toxicity of runoff and soil samples to the following aquatic and soil organisms: *D. magna* (48 h-survival, LC_{50} of 0.8 mg L^{-1}, and 21 days-reproduction, EC_{50} of 0.49 mg L^{-1}), *C. riparius* (10 days growth, EC_{50}: 92.5 mg L^{-1}), the soil invertebrates *Folsomia candida* (28 days reproduction, EC_{50}: 19.9 mg kg^{-1}) and *Enchytraeus crypticus* (28 days reproduction, E_{50}: 30.3 mg kg^{-1}), and the nematode *Caenorhabditis elegans* (72 h-reproduction, EC_{20}: around 1.4 mg L^{-1}).

A multi-parameter approach to assess the toxicity of PYR in a probable global change scenario has been used by Müller et al. [22], Seeland et al. [23] and Scherer et al. [25], based on the assumption that under climate change conditions warmer and more humid environments are expected, leading to conditions suitable for fungi development, thus an increase in the use of fungicide [22]. Therefore, these authors evaluated if the toxicity of PYR for invertebrate species (*C. riparius, D. magna, D. pulex, P. acuta*) alters with increasing temperature. Lethal PYR toxicity to *C. riparius* increased when combined with increasing temperature [23]. The loss of genetic diversity in *C. riparius* cohorts when exposed to PYR (2 mg L^{-1} = NOEC/2 of

PYR for reproduction) for multiple generations also depended on the thermal regime; genetic diversity became reduced by approximately %20 under thermal simulation of a typical cold or warm year in 2005–1990 and by %42 under a suboptimal temperature regime expected for a warm year in Europe in 2080–2050 under climate change conditions [22]. Likewise at suboptimal temperature conditions, the thermophil snail *P. acuta* presented higher susceptibility to toxic effects of PYR [24]. However, other studies indicate that PYR toxicity is highest at current optimal temperature regimes. The release of neonates from adult *D. magna* exposed to 0.5 mg L^{-1} (NOEC of PYR for reproduction) for multiple generations under dynamic temperature scenarios was most affected by PYR at a favorable temperature range (20 to 27 °C) [23]. In *D. pulex*, the inhibition of reproduction was not observed at suboptimal 15 °C, but at optimal 20 °C and 25 °C [25]. Interestingly, those PYR effects vanished in presence of kairomones from the predator *Chaoborus flavicans*.

5. Role of Pyrimethanil as Environmental Disturber: Avoidance Assays

It has been hypothesized that contaminants can act as toxicants as well as habitat disruptors. The former role is characterized by directly measuring acute or chronic responses in organisms, while their role as habitat disruptor is directly linked to effects on habitats, reducing their quality and triggering avoidance before toxic effects are detected. The latter effect is particularly important given that concentrations at which it might occur could be considered non-risky as no toxic effect at the individual level would be usually observed [43, 44]. Habitat disturbance caused by contamination as a result of agricultural activities may, therefore, be considered an additional factor that increases the threat of local population decline [26, 45].

Given the above, a new approach based on avoidance as an endpoint and using a non-forced exposure system has been proposed to assess the role of contaminants as environmental disturbers. This approach considers that contaminants as environmental disturbers can change the community structure with no direct toxic effect on organisms as they may be able to detect and avoid contaminants [26, 43–45]. The exposure system used here creates a contamination gradient in which organisms can freely move across different levels of contamination and choose the less contaminated zone. A few studies have tested this methodology and proved that contamination levels lower than those considered potentially dangerous for organisms can trigger avoidance response by many aquatic organisms [26, 43, 44]. As a consequence, an ecosystem can suffer structural changes as individuals able to detect contamination move towards less contaminated zones [44–46].

Avoidance tests in non-forced exposure systems (Figure 3) in which a PYR gradient was simulated have been performed with fries of *Danio rerio* [26] and tadpoles of amphibians *Lithobates catesbeianus* and *Leptodactylus latrans* [27].

Data obtained from these experiments showed that the spatial distribution of the three species was influenced by the presence of PYR. Almost all organisms of all three species avoided PYR when the concentration was around 1 mg L^{-1}. The results indicate, therefore, that organisms could react by avoiding a given environment before deleterious toxic effects set in and are detected via traditional acute endpoints. Based on the organisms' accumulated frequency along the system, the median preferred concentration, PC_{50} – concentration/dilution above or below which was preferred by %50 of organisms – and the PC_{25} and PC_{75} were calculated (Figure 3). For the three species, PC_{25}, PC_{50}, and PC_{75} were very similar and in general concentrations higher than 0.5 mg L^{-1} were avoided by %50 of the population of tadpoles and fries.

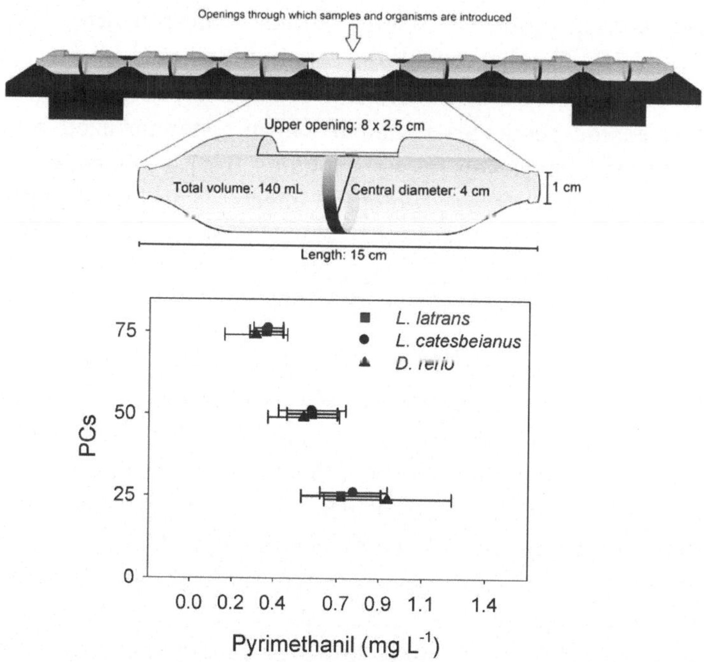

Figure 3. Schematic representation of the multi-compartmented non-forced system used for simulating a pyrimethanil gradient during avoidance assays (upper) and preferred concentration by 25, 50, and 75% (PC_{25}, PC_{50}, and PC_{75}) of tadpoles of two species of amphibians (Leptodactylus latrans and Lithobates catesbeianus) and of fries of Danio rerio exposed to a pyrimethanil gradient for 4 h (lower).

According to these findings, we emphasize the importance of taking into account the risk of the presence of plant protection products in the environment, even at non-lethal concentrations, due to their potential to trigger emigration. The presence of PYR can be a decisive factor in the habitat selection process of many species, such as shown in Figure 3. The disturbing effect of contaminants on ecosystems can be comparable to the loss and fragmentation of habitats [47, 48]. Habitats with reduced quality due to presence of contaminants probably may support a smaller population as well as lose the capacity to serve as sink habitats for surrounding populations [48]. Since avoidance experiments can provide information about contamination-driven habitat selection, the use of non-forced exposure systems is therefore encouraged in environmental risk assessment with agrochemicals.

6. Final Remarks

Undoubtedly, agrochemicals are potentially dangerous for aquatic organisms. The PYR concentration causing 50% reduced offspring in the most vulnerable aquatic species *D. pulex* (OECD model organism) is almost identical to the predicted environmental concentration of PYR in surface waters nearby apple orchards (0.089 mg L^{-1}, [17]). Thus, zooplankton communities may be at risk in case of expected PYR runoff into surface waters. If considering a risk safety factor of ten traditionally used for pesticides tested in chronic standardized bioassays with species from three trophic levels (algae, daphnids, fish), a PYR concentration of 0.9 mg L^{-1} should not induce adverse biological effects. Experiments with non-model species imply, however, that in particular Physidae are at risk at <0.9 mg L^{-1}. This result may recommend for the inclusion of different mollusk species in ecological risk assessment programs [24].

The similarly PYR-sensitive diatom *G. parviculum* [unpublished data] serves as important food source for grazers such as Physidae and one may therefore assume that indirect effects on the food web will appear in PYR-contaminated habitats in addition to the direct growth inhibition of the diatom, a result derived from single-species tests. Other indirect effects of 0.9 mg L^{-1} PYR may arise from the %50 avoidance behavior of the frogs *L. catesbeianus* and *L. latrans* observed after 12 h of exposure [27]. At a similar concentration range, the fish *D. rerio* avoids PYR contaminated freshwaters after 4 h (AC_{50} = 1.1 mg L^{-1}) [26]. The migration of top predators from certain habitats could however provide improved conditions for predated species (top-down effect), in particular for macroinvertebrates and algae being more PYR tolerant.

The available information provides evidences that PYR is potentially toxic for many aquatic species, affecting survival, reproduction, feeding, growth, and that it can disturb the environmental quality with no direct effect at the individual level by inducing organisms to migrate to less impacted areas. Although the amount of relevant information on the toxic potential of PYR on several species is increasing, little information is available on how the presence of PYR (and "inert compounds") can disturb broader environmental processes: chemical balance, direct effects on primary producers and consumers, changes in structure and functioning of the community and alterations in dispersion patterns. Further studies on the probable risk due to spray-drift, leaching, run-off, or accidental spills have to be encouraged. Presently, outdoor mesocosm studies taking into account different species, endpoints and exposure types in a more complex and relevant approach by using mesocosm experiments are ongoing. Given that the behavior and effects of PYR could vary between different climate conditions, the latter experiments are being performed across different climatic regions, from tropical to South- and North-temperate. Under these three environments, chemical dynamics of PYR in water and sediment are being followed for at least a 1 year period together with the monitoring of the complex local community, individual sub-lethal effects, changes in biodiversity and implications in ecological succession. The compilation of that information could help to understand the possible role that PYR plays environmental disturber for aquatic biota.

References

1. Schreinemachers P, Tipraqsa P. Agricultural pesticides and land use intensification in high, middle and low income countries. Food Policy 2013;37 616–626.
2. The World Bank. Agriculture for Development. World Development Report 2008. The World Bank, Washington, DC; 2008.
3. Hirooka T, Ishii H. Chemical control of plant diseases. Journal of General Plant Pathology 2013;79 390–401.
4. Grube A, Donaldson D, Kiely T, Wu L. Pesticides Industry Sales and Usage: 2006 and 2007 Market Estimates. United States Environmental Protection Agency, EPA 733-R-11-001, Office of Chemical Safety and Pollution Prevention (7503P), USA 34 pp; 2011.
5. Reilly TJ, Smalling KL, Orlando JL, Kuivila KM. Occurrence of boscalid and other selected fungicides in surface water and groundwater in three targeted use areas in the United States. Chemosphere 2012;89 228–234.
6. Petit A-N, Fontaine F, Vatsa P, Clément C, Vaillant-Gaveau. Fungicide impacts on photosynthesis in crop plants. Photosynthesis Research 2012;111 315–326.
7. Verdisson S, Couderchet M, Vernet G. Effects of procymidone, fludioxonil and pyrimethanil on to non-target aquatic plants. Chemosphere 2012;44 467–474.
8. Dewez D, Geoffroy L, Vernet G, Popovic R. Determination of photosynthetic and enzymatic biomarkers sensitivity used to evaluate toxic effects of copper and fludioxonil in alga Scenedesmus obliquus. Aquatic Toxicology 2005;74 150–159.
9. Navarro S, Barba A, Navarro G, Vela N, Oliva J. Multiresidue method for the rapid determination - in grape, must and wine - fungicides frequently used on vineyards. Journal of Chromatography A 2000;882 221–229.
10. Moyano C, Gómez V, Melgarejo P. Resistance to pyrimethanil and other fungicides in Botrytis cinerea populations collected on vegetable crops in Spain. Journal of Phytopathology 2004;152 484–490.

11. Anfossi L, Sales P, Vanni A. Degradation of anilinopyrimidine fungicides photoinduced by iron (III)-polycarboxylate complexes. Pest Management Science 2006;62 872–879.
12. Mosleh YY, Mofeed J, Afifi M, Almaghrabi OA. Biological effects of pyrimethinal on aquatic worms (Tubifex tubifex) under laboratory conditions. Bulletin of Environmental Contamination and Toxicology 2014;92 85–89.
13. Vázquez D, Panozzo M, Almirón N, Bello F, Burdyn L, Garrán S. Characterization of sensitivity of grove and packing house isolates of Penicillium digitatum to pyrimethanil. Postharvest Biology and Technology 2014;98 1–6.
14. Shinn C, Delello-Schneider D, Mendes LB, Sanchez AL, Müller R, Espíndola ELG, Araújo CVM. Immediate and mid-term effects of pyrimethanil toxicity on microalgae by simulating an episodic contamination. Chemosphere 2015;120 407–413.
15. FAO/WHO. Pesticide residues in food. Part II — toxicological. Joint FAO/WHO meeting on pesticide residues, Switzerland; 529 pp; 2007.
16. EFSA – European Food Safety Authority. Review of the maximum residue levels (MRLs) for pyrimethanil according to Article 12 of Regulation (EC) No 396/2005. EFSA Journal 2011;11 2454.
17. EFSA – European Food Safety Authority. Conclusion regarding the peer review of the pesticide risk assessment of the active substance pyrimethanil. EFSA Scientific Report 2006;61 1–70.
18. Smilanick JL, Mansour MF, Mlikota Gabler F, Goodwine WR. The effectiveness of pyrimethanil to inhibit germination of Penicillium digitatum and to control citrus green mold after harvest. Postharvest Biology and Technology 2006;42 75–85.
19. Sugar D, Basile SR. Timing and sequence of postharvest fungicide and biocontrol agent applications for control of pear decay. Postharvest Biology and Technology 2008;49 107–112.
20. Sirtori C, Zapata A, Malato S, Agüera A. Formation of chlorinated by-products during photo-Fenton degradation of pyrimethanil under saline conditions. Influence on toxicity and biodegradability. Journal of Hazardous Materials 2012;217–218 217–223
21. Dosnon-Olette R, Couderchet M, Eullaffroy P. Phytoremediation of fungicides by aquatic macrophytes: Toxicity and removal rate. Ecotoxicology and Environmental Safety 2009;72 2096–2101.
22. Müller R, Seeland A, Jagodzinski LS, Diogo JB, Nowak C, Oehlmann J. Simulated climate change conditions unveil the toxic potential of the fungicide pyrimethanil on the midge Chironomus riparius: a multigeneration experiment. Ecology and Evolution 2012;2 196–210.
23. Seeland A, Oehlmann J, Müller R. Aquatic toxicity of the fungicide pyrimethanil: effect profile under optimal and thermal stress conditions. Environmental Pollution 2012;168 161–169.

24. Seeland A, Albrand J, Oehlmann J, Müller R. Life stage-specific effects of the fungicide pyrimethanil and temperature on the snail Physella acuta (Draparnaud, 1805) disclose the pitfalls for the aquatic risk assessment under global climate change. Environmental Pollution 2013;174 1–9.

25. Scherer C, Seeland A, Oehlmann J, Müller R. Interactive effects of xenobiotic, abiotic and biotic stressors on Daphnia pulex – results from a multiple stressor experiment with a fractional multifactorial design. Aquatic Toxicology 2013;138–139 105–115.

26. Araújo CVM, Shinn C, Mendes LB, Delello-Schneider D, Sanchez AL, Espíndola ELG. Avoidance response of Danio rerio to a fungicide in a linear contamination gradient. Science of the Total Environment 2014;484 36–42.

27. Araújo CVM, Shinn C, Vasconcelos AM, Ribeiro R, Espíndola ELG. Preference and avoidance responses by tadpoles: the fungicide pyrimethanil as a habitat disturber. Ecotoxicology 2014;23 851–860.

28. PPDB. The Pesticide Properties Database (PPDB) Developed by the Agriculture & Environment Research Unit (AERU), University of Hertfordshire, Funded by UK National Sources and the EU-funded FOOTPRINT Project (FP6-SSP-022704); 2009.

29. Ma J, Wang S, Wang P, Ma L, Chen X, Xu R. Toxicity assessment of 40 herbicides to the green alga Raphidocelis subcapitata. Ecotoxicology and Environmental Safety 2006;64 456–462.

30. Lewis MA. Use of freshwater plants for phytotoxicity testing: a review. Environmental Pollution 1995;87 319–336.

31. Ma J. Differential sensitivity to 30 herbicides among population of two green algae Scenedesmus obliquus and Chlorella pyrenoidosa. Bulletin of Environmental Contamination and Toxicology 2002;68 275–281.

32. OECD – Organisation for Economic Co-operation and Development. OECD guidelines for the testing of chemicals. Lemna sp. growth inhibition test. Guideline 221. Paris, France; 2006.

33. Ma J, Lin F, Wang S, Xu L. Toxicity of 21 herbicides to the green alga Scenedesmus quadricauda. Bulletin of Environmental Contamination and Toxicology 2003;71 594–601.

34. Gómez de Barreda Ferraz D, Sabater C, Carrasco JM. Effects of propanil, tebufenozide and mefenacet on growth of four freshwater species of phytoplankton: a microplate bioassay. Chemosphere 2004;56 315–320.

35. Liu S-S. Wang C-L, Zhang J, Zhu X-W, Li W-Y. Combined toxicity of pesticide mixtures on green algae and photobacteria. Ecotoxicology and Environmental Safety 2013;95 98–103.

36. Feiler U, Ratte M, Arts G, Bazin C, Brauer F, Casado C, Dören L, Eklund B, Gilberg D, Grote M, Gonsior G, Hafner C, Kopf W, Lemnitzer B, Liedtke A, Matthias U, Okos E, Pandard P, Scheerbaum D, Schmitt-Jansen M, Stewart K, Teodorovic I, Wenzel A, Pluta H-J. Inter-laboratory trial of a standardized sediment contact test with the

aquatic plant Myriophyllum aquaticum (ISO 16191). Environmental Toxicology and Chemistry 2014;33 662–670.

37. van Leeuwen L, Vonk JW. Environmental risk limits for pyrimethanil. RIVM National Institute for Public Health and the Environment, Letter report 601716010/2008, The Netherlands, 20p; 2008.
38. McWilliam RA, Baird DJ. Postexposure feeding depression: a new toxicity endpoint for use in laboratory studies with Daphnia magna. Environmental Toxicology and Chemistry 2002;21 1198–1205.
39. McWilliam RA, Baird DJ. Application of postexposure feeding depression bioassays with Daphnia magna for assessment of toxic effluents in rivers. Environmental Toxicology and Chemistry 2002;21 1462–1468.
40. Lopes I, Moreira-Santos M, da Silva EM, Sousa JP, Guilhermino L, Soares AMVM, Ribeiro R. In situ assays with tropical cladocerans to evaluate edge-of-field pesticide runoff toxicity. Chemosphere 2007;67 2250–2256.
41. Satapornvanit K, Baird DJ, Little DC. Laboratory toxicity test and post-exposure feeding inhibition using the giant freshwater prawn Macrobrachium rosenbergii. Chemosphere 2009;74 1209–1215.
42. Gil FN, Moreira-Santos M, Chelinho S, Pereira C, Feliciano JR, Leitão JH, Sousa JP, Ribeiro R, Viegas CA. Suitability of a Saccharomyces cerevisiae-based assay to assess the toxicity of pyrimethanil sprayed soils via surface runoff: comparison with standard aquatic and soil toxicity assays. Science of the Total Environment 2015;505 161–171.
43. Lopes I, Baird DJ, Ribeiro R. Avoidance of copper contamination by field populations of Daphnia longispina. Environmental Toxicology and Chemistry 2004;23 1702–1708.
44. Moreira-Santos M, Donato C, Lopes I, Ribeiro R. Avoidance tests with small fish: determination of the median avoidance concentration and of the lowest-observed-effect gradient. Environmental Toxicology and Chemistry 2008;27 1575–1582.
45. Rosa R, Materatski P, Moreira-Santos M, Sousa JP, Ribeiro R. A scaled-up system to evaluate zooplankton spatial avoidance and population immediate decline concentration. Environmental Toxicology and Chemistry 2012;31, 1301–1305
46. Araújo CVM, Shinn C, Moreira-Santos M, Lopes I, Espíndola ELG, Ribeiro R. Copper-driven avoidance and mortality in temperate and tropical tadpoles. Aquatic Toxicology 2014;146 70–75.
47. Ribeiro R, Lopes I. Contaminant driven genetic erosion and associated hypotheses on alleles loss, reduced population growth rate and increased susceptibility to future stressors - an essay. Ecotoxicology 2013;22 889–899.
48. Wilson JD, Hopkins WA. Evaluating the effects of anthropogenic stressors on source-sink dynamics in pond-breeding amphibians. Conservation Biology 2013;27 595–604.

7

Toxicity of Agrochemicals on Freshwater Invertebrates — A Short Review

1. Introduction

The need to provide enough food to the growing worldwide population and control pests are some factors that have led to the application of agrochemicals (pesticides) on agricultural areas to protect and increase the crop production [1]. Despite the advantages offered by pesticides, these substances can turn into an environmental concern since they can leave their action point mainly by surface water runoff and reach water reservoirs, which could alter the aquatic environment and pose a threat to human health [2-4]. The majority of these chemicals have a synthetic basis, and different categories have been established to classify them depending on their chemical structure. Some of the most representative agrochemicals with great ecological impact are organochlorine hydrocarbons (DDT), organophosphates (parathion and diazinon), carbamates (carbaryl and methiocarb) and pyrethroids (deltamethrin), where the first group is characterized by its stability in the environment after being released. Pollution of freshwater ecosystems with these chemicals is well known and has been reported for several regions worldwide and it represents a problem of consideration for the preservation of the aquatic environment [5-8]. All these pesticides act by altering the organism's nervous system [6].

In this context, scientists have worked to develop and standardize protocols to evaluate the toxic effects of a wide variety of pollutants on certain living organisms known as "sentinel organisms" or "bioindicators" [9]. Bioassays are toxicity studies that can be performed with organisms that represent an important component of ecosystems and are able to respond to xenobiotics, and therefore, bioassays may be used to predict "safe levels" of toxicants in the environment. Among bioindicators, freshwater invertebrates are used frequently due to their importance as primary consumers of algae and herbivores representing a key link in trophic webs [10-12]. Moreover, some aspects like a) abundance, b) wide distribution, c) maintenance and easy culture in the laboratory, d) genetic stability and e) sensitivity are considered to select test organisms [13, 14].

Standard toxicity tests are usually performed with a single species to assess the toxicity in water samples and different endpoints can be evaluated, such as motility, reproduction and enzymatic inhibition. The endpoint "motility", usually corresponds to a short-term (acute) toxicity assay and represents the concentration of chemicals that reduces the motility to 50% of the animals after 24 or 48 h exposure and the result is expressed as EC_{50}. This assay also can be interpreted as the lethal concentration for 50 percent of individuals (LC_{50}). For the long-term (chronic) tests, behavioral changes (grazing and filtration rates, phototaxis and survival) and reproduction assays can be conducted. The reproduction assay evaluates the effects on reproduction typically after 21 days of exposure and is represented by EC_{50}. This parameter estimates the concentration that inhibits 50% of reproductive effort [15-19].

Within the chronic test category, ecotoxicologists have used another approach to evaluate toxicity known as "sub-lethal effects tests" by estimating variations in biochemical or physiological components (biomarkers). Their importance is based on their capability to indicate damage to the organism following exposure to concentrations of contaminants that are not acutely toxic. Some examples are the enzymatic inhibition and genomic responses (genotoxicity) that indicate disturbances occurring at the sub-organismal level [20-22].

Measurement of different endpoints can provide valuable toxicological information to derive water quality criteria for the safe release of compounds into aquatic bodies [13, 23]. This chapter focuses mainly on those studies performed with freshwater invertebrates that are representative for comparison purposes according to their availability in literature.

2. Freshwater Invertebrates as Sentinel Organisms

2.1. Cladocerans

In general terms, *Daphnia magna* (figure 1) and *Ceriodaphnia dubia* represent the main daphnids (Class Crustacea, order Cladocera, family Daphniidae) used as bioindicators. Moreover, *Daphnia carinata* and *Daphnia galeata* have also been tested and have been included in the chapter. These organisms known as "water fleas" have filter-feeding habits and are ubiquitous species in temperate freshwater bodies. Usually, daphnids reproduce by an asexual reproduction mechanism called "parthenogenesis" [15, 24, 25].

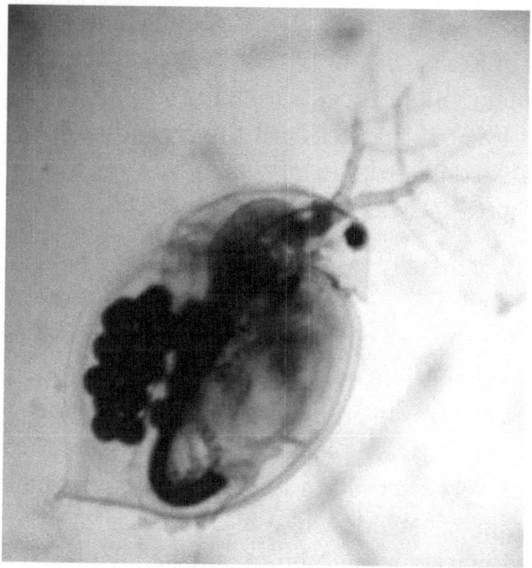

Figure 1. D. magna (Female individual. Photograph taken under stereomicroscope (January, 2015) by Doctor Gustavo E. Santos Medrano, Chemistry department, Universidad Autónoma de Aguascalientes)

2.1.1. Daphnids as bioindicators

2.1.1.1. Daphnia magna

Sánchez *et al.* (1999) [26] performed a two-generation reproduction test (chronic assay: 21-day life study) to assess the effects of diazinon at different concentrations ranging from 0.05 to 1.0 ng/L. This agrochemical is an organophosphorus pesticide used to fight leaf-eating insects. The parameters evaluated in the parental organisms (F_0) as in the offspring (first and third brood) were size, survival and fecundity. Between the main findings, a remarkable decrease in longevity, number of individuals per female, brood size and number of broods per female in F_0 was noted as the pesticide concentration was increased. These parameters did not decrease dramatically in the offspring as compared with control and individuals showed a higher reproduction rate than their parental mothers. Moreover, fecundity and growth from first and third brood did not recover completely. A 24h LC_{50} of 0.86 μg/L was obtained.

Bettinetti *et al.* (2013) [27] monitored reproduction (21-day assay) and survival as toxicity endpoints in daphnids exposed to pp´-DDE (pp´- Dichlorodiphenyl – dichloroethylene), a more stable metabolite of the organochloride p,p´-DDT (dichlorodiphenyl-trichloroethane) through diet and to different concentrations in water. For the treatment with contaminated algae, a decrease of %83.3 was registered in neonates production for mothers exposed to the highest dose (795.6 pp´DDE ng/per organism) at day 21 and were shorter in length compared to control group. A reduction in grazing activity (i.e. ingestion activity) was also observed in mothers with 24 ng/mg (dry weight basis). Exposure to pp′-DDE in water showed that the maximum metabolite concentration used (6 μg/L) increased the mother's long-term mortality by 50% and reduced the fecundity of the surviving mothers by 32.4%. A 48h IC_{50} (concentration at which %50 of the individuals presented immobilization) of 5.08 μg/L was estimated.

Another modality of toxicity test corresponds to the behavioral response under toxicant exposure. Martins *et al.* (2007) [28] evaluated the phototactic behavior "phototaxis" in *D. magna*, which consists of individuals' movement towards or away from a source light as a possible outcome of different natural phenomena (finding prey, reproduction) and stress (predator avoidance, photochemical damage) [29-31]. In this study, strong positive phototactism (movement toward a light source) clones were used and were exposed to 11 chemicals commonly found in freshwater environments, including pesticides. It was observed that the fungicide "Thiram" (carbamate) reduced markedly the phototactic behavior and was detected within 0.25 h of exposure at a concentration of 9.38 μg/L. Furthermore, carbamate was detected at a lower concentration than the 48h LC_{50} (210 μg/L) reported for the bioindicator.

Toxic effects of pesticides also have been evaluated considering food availability. Pereira and Golcalves (2007) [32] evaluated acute and chronic toxicity of methomyl (carbamate) to different daphnid species, including *Daphnia magna*, under different food level regimes. *D. magna* showed the greater resistance to the toxicant and presented a 48h EC_{50} (concentration at which %50 of the individuals presented immobilization) of 24.17 μg/L. For the chronic toxicity tests, a decrease in reproduction was observed as methomyl concentrations were increased and sensitivity was greater at low food levels.

Sensitivity between same species has been tested by Toumi *et al.* (2013) [33], who studied the toxic effects of the pyrethroid "deltamethrin" on two daphnid strains provided from different laboratories. The results of acute toxicity tests (immobilization) revealed a 48h EC_{50} of 0.32 and 0.63 μg/L, for strain 1 and 2, respectively. In terms of chronic toxicity (survival and reproduction), deltamethrin induced significant effects and embryo deformities were found, which, according to the authors, gives certain evidence that this agrochemical could have endocrine disruptive effects. Nevertheless, it is worth mentioning that differences in sensitivity for the two strains were observed.

2.1.1.2. Ceriodaphnia dubia

Shen *et al.*, (2012) [34] have reported some toxicological data using *C. dubia* after conducting acute and chronic assays for two pyrethroids: deltamethrin and α-cypermethrin. Results showed lethal toxicity on the cladoceran and 48h LC_{50} for immobilization of 0.06 and 0.84 μg/L were recorded for each agrochemical, respectively. For the 8-day chronic assays, survival and reproduction endpoints were evaluated, thus obtaining an EC_{50} of 116 and 34.7 ng/L for deltamethrin and 209 and 97.8 ng/L for α-cypermethrin, respectively.

Metabolic activation of pesticides via cytochrome P450 (a protein superfamily involved in the metabolism of xenobiotics and endogenous compounds) has been tested by El-Merhibi *et al.* (2004) [35] by studying the toxicity of the organophosphorus chlorpyrifos in the presence or absence of the inhibitor/inducer of cytochrome P450 "piperonyl butoxide (PBO)". Among the main findings, a 48h LC_{50} for mortality was estimated at 0.05 µg/L in the absence of PBO. Moreover, a reduction in acute toxicity was evident in the presence of piperonyl butoxide in response to increasing concentrations of this chemical.

2.1.1.3. Daphnia carinata

The Australian native species *D. carinata* has been tested by Cáceres *et al.* (2007) [36] to evaluate the acute toxicity of chlorpyrifos and its metabolite 3,5,6-trichloropyridinol (TCP) separately and in combination tests using a cladoceran media and river water. A greater toxic effect was observed in the cladoceran media by TCP (48 LC_{50}: 0.20 µg/L) but showed no toxicity in natural water, whereas the parental compound exhibited a 48h LC_{50} for river water and cladoceran media of 0.3 and 0.24 µg/L, respectively. The absence of toxicity by the metabolite in river water was attributed to the microbial activity that led to its degradation. The toxicity assessment using both chemicals at a concentration of 0.12 µg/L did not affect the survival of the species in natural water but reflected an additive effect and caused 72% mortality in the artificial media. According to the researchers, this finding suggests that joint toxicity could behave in a different manner when chemicals are in the environment and differences between the media composition should be considered.

Acute and chronic toxicity of chlorpyrifos was evaluated by Zalizniak and Nugegoda (2006) [37] using three successive daphnid generations. For the lethal toxicity, a 48h LC_{50} was estimated for parent generation (0.5 µg/L). In long-term toxicity assays (21-day survival), fecundity, time to the first brood and female size were monitored. The number of offspring per female in parent individuals was significantly reduced. The main endpoints altered in the first generation were survival and fecundity, whereas the time to the first brood and an indication of hormesis (response stimulation and inhibition at low and high concentrations, respectively) were evident in the second generation. Moreover, the lowest concentration tested (0.005 µg/L) yielded the lowest number of offspring per female. For the third generation, daphnids showed a remarkable sensitivity at low concentrations of chlorpyrifos (0.025 µg/L).

2.1.1.4. Daphnia galeata

Some researchers have studied the effects of chlorpyrifos using *Daphnia galeata* as bioindicator. van Wijngaarden *et al.* (1993) [38] monitored immobility and mortality as acute toxicity endpoints and a 48h EC_{50} of 0.3 μg/L was recorded. van den Brink *et al.* (1995) [39] evaluated chronic toxicity (24 days) simulating an indoor microcosm, including different zooplankton (*D. galeata*), phytoplankton and macro-invertebrate species under low concentration exposure (0.1 μg/L). Daphnid population reduced to zero within the first week of the test, nevertheless, stress factors like predation could have influenced this finding. Moreover, the agrochemical affected zooplankton species, which resulted also in community alterations.

2.1.1.5. Biomarkers in *Daphnia magna*

In relation to agrochemical toxicity assessment using biomarkers, Guilhermino *et al.* (1996) [40] proposed inhibition of the enzyme acetylcholinesterase (AChE) as an acute toxicity indicator. This enzyme involved in neural transmission has been monitored under carbamates and organophosphates exposure. Liu *et al.* (2012) [41], conducted both an acute and a sub-lethal study (21 days) with chlorpyrifos. In acute bioassays, 48h EC_{50} for immobilization was (7.12 μg/L). Moreover, in the chronic test, a recovery on AChE activity was noted after the second day of exposure, this probably because *D. magna* developed adaptive mechanism(s) to mitigate stress.

A genotoxicity study was conducted by Pereira *et al.* (2010) [42] to analyze gene transcription after acute exposure (48 hours) of third to fifth brood juvenile daphnids to the insecticide methomyl at a concentration of 10.5 μg/L. It was found that the agrochemical was able to induce several genes and affect neuronal transmission. mRNAs of a digestive enzyme (α-amylase) and diverse lipoproteins were up-regulated, which according to the authors followed the need of carbohydrate breakdown for energy production and was an effort to maintain homeostasis under toxicant stress. mRNAs of genes involved in defense mechanisms (galactose-binding C-type lectins, cystatins and ferritins) were also up-regulated. As mentioned by the authors, lectins and cystatins play an important role in the general immune response as they participate in hemolymph coagulation and ferritin expression might indicate oxidative stress. Even when stress responses were evident, a strong evidence for expression responses related exclusively to genes associated with the pesticide target site was not found.

An assay to elucidate toxicity mechanisms of carbamates using a biomarker (AChE) was implemented by Jeon *et al.* (2013) [43] by exposing *Daphnia magna* to carbaryl (carbamate). Results were compared to the USEPA ECOTOX database for the organophosphorus irreversible AChE inhibitor "diazinon". For the carbamate, an IC_{50} for *in vitro* AChE activity and a 48h EC_{50} value for immobilization were obtained (0.56 μM and 63 nM, respectively). The latest parameter was greater than that proposed by USEPA ECOTOX for diazinon (3.0 nM). Low toxicity was observed by carbaryl acting through a reversible inhibition mechanism. This finding was attributed to the instability of the AChE-carbaryl complex where a lower hydrophobicity of this chemical is possible. A declining energy reserve was also identified since lipid and glycogen reservoirs decreased with an increase of protein content.

In another study, Toumi *et al.* (2015) [44] estimated AChE activity in 3 *Daphnia magna* strains (strains 1 and 3 already identified; clonal identification of strain 2 remains unknown) after deltamethrin exposure, a pyrethroid insecticide, reporting a significant decrease on the enzyme activity in 2 strains. Variation in sensitivity between strains was observed and different 48h IC_{50} (concentration at which %50 of AChE was inhibited) and 48h EC_{50} values were obtained for immobilization endpoint. Thus, the strain 1 registered the lower EC_{50} (0.32 µg/L) and strain 3 the lower IC_{50} (0.016 µg/L). For strain 2, an EC_{50} of 0.63 µg/L and IC_{50} of 0.018 µg/L were registered. Lowest Observed Effect Concentration (LOEC) for each strain was also estimated (Strain1: 80.6 ng/L; Strains 2 and 3: 20.1 ng/L). According to the authors, the interclonal variability observed in these results can lead to propose AChE as a biomarker of susceptibility (the response is specific for each strain) for deltamethrin exposure in D. magna.

Barata *et al.* (2004) [45] also assessed inhibition of AChE and carboxylesterase (CbE) after pulse exposures of *D. magna* to three pesticides in the presence and absence of triphenyl phosphate and -2(O-cresyl)4-H3,2-,1-benzodioxaphosphorin-2-oxide (CBDP) CbE inhibitors. A 24h LC_{50} of 12.38 ,1.28 and 762.93 nM was estimated for chlorpyrifos, malathion and carbofuran (carbamate), respectively. Both enzymes in the treatments with the carbamate were inhibited in less than 2 hours as this toxicant does not need to be activated by cytochrome P450 enzymes. Moreover, results showed higher sensitivity of CbE (lower concentration inhibition) against organophosphorus pesticides, which according to the authors was due to a higher enzyme affinity to the oxons formed (malaoxon and chlorpyrifos-oxon). It was suggested that this finding could be involved in conferring protection to AChE by sequestering available pesticides. Besides, the two CbE inhibitors increased mortality response as more agrochemicals availability was present for AChE. β-esterase activity in individuals exposed to carbofuran reached an activity similar to control level in less than 12h, whereas in daphnids exposed to organophosphates both enzymes resumed their activity between 24 and 96h to achieve 50% and almost complete recovery levels, respectively.

Some digestive enzymes have been used as biomarkers. De Coen *et al.* (1998) [46] monitored ingestion activity using fluorescent labeled latex microbeads and the digestive enzymes "esterase, trypsin and β-galactosidase" under the exposure of pentachlorophenol (PCP) and lindane, both organochloride pesticides. Considerable reduction in esterase (greater than 50%) and ingestion (90min EC_{50}: 0.3 mg/L) activities were induced by lindane. Moreover, a drastic decrease in ingestion activity was noticed compared to digestive enzyme inhibition. According to the authors, this was probably due to an energy-saving strategy, since ingestion demands more activity, and therefore, daphnids could maintain a certain food assimilation efficiency. 24h EC_{50} values for immobilization were estimated (PCP: 0.7 mg/L; Lindane: 1.8 mg/L).

Another aspect of interest in ecotoxicology is that some pesticide metabolites can exhibit more toxicity than their parental compounds as it has been reported previously by Belden and Lydy (2000) [47]. Guilhermino *et al.* (1996) [48] tested AChE activity with *D. magna* under the organophosphorus pesticide "parathion" as parental chemical and its metabolite "paraoxon" and observed such phenomena as the EC_{50} values from *in vivo* AChE inhibition and acute toxicity (48h EC_{50}) for parathion (2.4 and 2.2 μg/L, respectively) were higher than those obtained for paraoxon: 0.2 μg/L for both *in vivo* AChE inhibition and acute assay.

2.2. Rotifers

Within the phylum Rotifera, rotifers of the genus *Brachionus* have been widely used as test organisms due their cosmopolitan distribution, ease of maintenance in the laboratory, rapid reproduction and availability of resting eggs (widely used in ecotoxicology) as these microscopic freshwater invertebrates play an important link in food webs by making biomass available for higher trophic levels [49-51]. For comparison purposes, *Brachionus calyciflorus* (figure 2), *Brachionus patulus* and one rotifer from the genus *Lecane* were included.

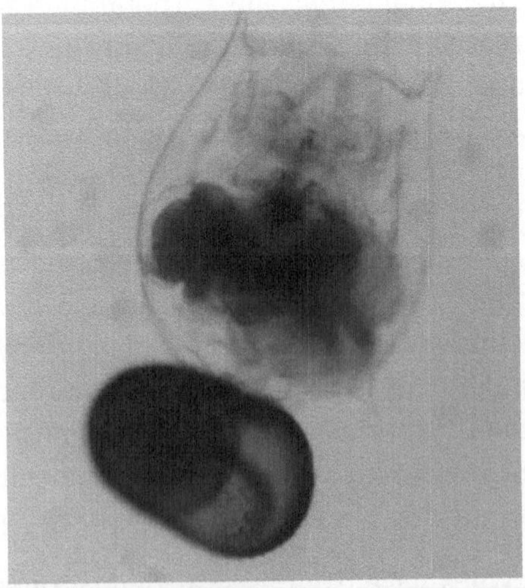

Figure 2. B. calyciflorus (Female individual. Photograph taken under optical microscope (January, 2015) by Doctor Gustavo E. Santos-Medrano, Chemistry department, Universidad Autónoma de Aguascalientes)

2.2.1. Brachionus calyciflorus

Fernández-Casalderrey *et al.* (1992) [52] studied the chronic effects of diazinon on *B. calyciflorus*. A 24h LC_{50} of 29.22 mg/L was estimated. Low concentrations of the agrochemical (range: 19–5 mg/L) were capable to reduce survival, fertility, life expectancy and reproduction with a dose-response relationship.

This researcher group also evaluated chronic toxicity of the organophosphorus methylparathion by feeding the rotifers with *Nannochloris oculata* and *Chlorela pyrenoidosa*. A 24h LC_{50} of 29.19 mg/L was estimated. Life expectancy and reproduction rate were reduced with increasing concentrations of the pesticide. Chronic exposure at 5 mg/L could lead to rotifer population extinction. Moreover, the decline was greater on those individuals fed with *Chlorela* (larger in size than *Nannochloris*) [53].

Ke *et al.* (2009) [54] studied the effects of three pesticides including chlorpyrifos on population growth and sexual reproduction. The results showed an induction on the mictic rate (sexual reproduction) since resting egg production was increased at concentrations between 0.1 and 100 μg/L. A similar response was found for the population growth as this parameter was increased but only at pesticide concentrations from 0.01 to 100 μg/L. Mortality was observed only at the highest treatment concentration (1000 μg/L) after 24 h exposure.

In another study, ingestion rate was proposed as a sub-lethal stress indicator by Juchelka and Snell (1994) [55] for this species. *B. calyciflorus* was exposed to different dilutions of PCP, diazinon and chlorpyrifos under short exposure scheme (30 min) and fluorescence intensity was measured to estimate the number of microspheres ingested in 5 min after exposure treatments. 24h LC_{50}, ingestion (NOEC: No observed effect concentration) and reproductive rate (48h NOEC) values of 1.2, 0.13 and 0.11 mg/L, 31, 20 and 8 mg/L and 12, 0.25 and 0.23 mg/L were estimated for PCP, diazinon and chlorpyrifos, respectively.

2.2.2. Brachionus patulus

In [56], the effects of different sub-lethal concentrations of DDT under high and low food levels on *Brachionus patulus* were assessed. Survival and fertility were higher for the high food level. Authors hypothesized that this finding might be the result of a healthier status as the individuals were better fed, thus acquiring more toxicant resistance. Moreover, a dose-response relationship was observed for all endpoints tested in both food regimes.

2.2.3. Lecane quadridentata

Lecane quadridentata (figure 3) is found in Mexican waters (Lake Chapala) and has been used by several authors to assess different toxicity endpoints against several toxicants including metals, organic compounds and pesticides [57-60].

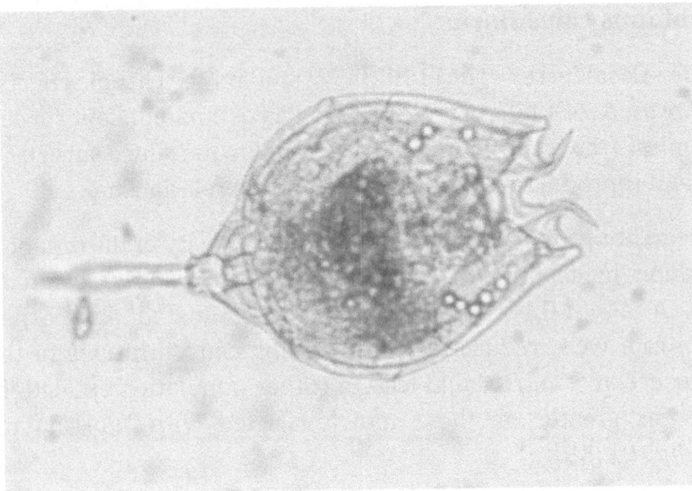

Figure 3. Lecane quadridentata (Female individual. Photograph taken under optical microscope (January, 2015) by Doctor Gustavo E. Santos-Medrano, Chemistry department, Universidad Autónoma de Aguascalientes)

In [61], a study was conducted to perform three toxicity tests "lethal (48h mortality), sub-lethal (inhibition of AChE activity) and chronic (5 day inhibition of the instantaneous growth rate) assays" using carbaryl and methylparathion. The carbaryl pesticide exhibited the higher chronic toxicity (EC50 2.22 mg/L) but greater lethal and sub-lethal toxicity was registered with the organophosphorus pesticide (9.4 mg/L for both bioassays). Moreover, the growth rate was more sensitive in comparison to the esterase activity and was proposed by the authors as a biomarker to assess the toxicity of anticholinesterase pesticides.

3. Relevance of Sensitivity of Cladocerans and Rotifers Against Pesticides

As mentioned earlier, freshwater invertebrates tend to be sensitive against pollutants and are key factors to maintain freshwater ecosystem quality, thus, preserving these organisms in their habitat is important to guarantee the entire water reservoir health. In the present chapter, different studies to assess the toxicity of four main classes of pesticides on freshwater invertebrates were reviewed.

According to the available data on literature, *Daphnia magna* has been widely used in toxicity assays with agrochemicals, nevertheless, it is not always the most sensitive species. In table 1, where additional references for comparison purposes were included (Additional References: [62- 69]), inter- and intraspecies differences on sensitivity are shown. For example, *Ceriodaphnia dubia* exhibited the greatest sensitivity in acute and chronic toxicity against lindane in comparison to *D. magna* and *Brachionus calyciflorus* and showed the lowest 24h LC50 for

PCP. Besides, this cladoceran showed the lowest 48h LC50 for the pyrethroids deltamethrin and cypermethrin and the organophosphorus pesticide diazinon. In some other cases (carbaryl and methylparathion), *D. magna* was more sensitive than the rest of the bioindicators. For the genus *Brachionus*, a marked tolerance was observed for the four agrochemicals considered. These differences could be attributed to the toxicant interaction with the test organism, differences in toxicity patterns of each agrochemical and to certain conditions used in the bioassay: type of assay, test duration, presence or absence of food and chemical concentration.

In this context, to obtain a more accurate toxicity estimation of water samples polluted with agrochemicals, different endpoints should be evaluated. For example, Pérez-Legaspi *et al.* (2010) included [61] acute and chronic toxicity assays and a biomarker to evaluate the toxicity of methylparathion and carbaryl on *Lecane quadridentata,* where the chronic test showed the higher sensitivity for both xenobiotics. In addition, the use of standardized protocols together with an additional test (using a different test organism) could help to obtain a more reliable conclusion about the actual condition of the freshwater system under evaluation. Also, mesocosm studies could help to estimate the effects of pesticides at the community level and predict long-term effects.

Additionally, when evaluating toxicity in water samples, some other considerations become important. In natural water bodies, it is likely that aquatic organisms are exposed to different agrochemicals and for longer periods, thus, phenomena such as synergism between pesticides and bioaccumulation (accumulation of substances in an organism) are possible and can aggravate the ecological impact by an increase in their toxicity as the pollutants could move through food chains and reach final consumers including the human being. When considering these two phenomena in ecotoxicological studies a more realistic and representative result can be obtained.

In relation to the water quality criteria for aquatic life protection (table 1), some values seem to be appropriate to protect aquatic organisms (lindane, PCP, DDE), however, in some cases there is no criteria available (methylparathion) and for other agrochemicals the recommended concentration could be not that protective, such as those proposed for malathion, chlorpyrifos, diazinon, cypermethrin, deltamethrin, carbaryl and DDE, as these values are close to the toxicity values registered for some endpoints. Pesticide toxicity on freshwater organisms is evident and the need for continuous generation of ecotoxicological data to protect aquatic life and human health still remains.

Table 1. Toxicological data for rotifers and cladocerans.

Agrochemical		**Species**	**Endpoint**				**Water quality criteria for aquatic life protection** (mg/L)
		Acute toxicity		**Chronic toxicity**	**Sub-lethal toxicity (biomarkers)**		
		24h LC_{50} (mortality) (mg/L)	48h LC_{50} (immobilization) (mg/L)	Reproduction (EC_{50}) (mg/L)	*In vivo* 48h IC_{50} (AChE) (mg/L)		
Organochloride pesticides	Lindane	*Dm*	1.8[p]	1.7[w]	0.34[w] (16 d)	-	2x10^{-4}**
		Cd	-	0.045[w] (7 day)	0.013[w] (7 day)	-	
		Bc	22.5[v]	-	-	-	
	PCP (sodium salt)	*Dm*	0.7[p]	0.35[w]	-	-	5x10^{-4}* (long term) 0.019^{+} at pH 7.8 (short term) 0.015^{+} at pH 7.8 (long term)
		Dc	-	1.3[w]	-	-	
		Cd	0.149[w]	-	-	-	
		Bc	1.2[c]	1.2[w]	0.27[w] (2 day)	-	
	DDE	*Dm*	-	5.08x10^{-3} [d]	-	-	3x10^{-5} **
Carbamates	Thiram	*Dm*	-	0.21[e]	-	-	2x10^{-4} **
	Methomyl			0.024[f]			3.5x10^{-3}**
	Carbaryl	*Dc*	0.1[w]	0.035[w]	-	-	3.3x10-3* (short term) 2x10-4* (long term)
		Cd	-	0.012[w]	8.6x10^{-3}[w] (7 day)	-	
		Dm	-	0.012[q]	-	-	
		Lq	-	13.72[r]	2.2[r] (5 day)	17.19[r](45 min)	

Pyrethroids	Deltamethrin	*Dm* (S1)	$9.4x10^{-3\ g}$	$3.2x10^{-4\ gi}$	-	$5.8x10^{-5\ i}$	$1x10^{-7**}$
		Dm (S2)	$8.86x10^{-3\ g}$	$6.3x10^{-4\ gi}$	-	$1.8x10^{-5\ i}$	
		Dm (S3)	-	$8.8x10^{-4i}$	-	$1.6x10^{-5\ i}$	
		Cd	$8.4x10^{-4\ h}$	$6x10^{-5\ h}$	$3.47x10^{-5\ h}$ (8 day)	-	
	Cypermethrin	*Cd* *Dm*	$2.5x10^{-3h}$ $2x10^{-3w}$	$8.4x10^{-4h}$ 89^{w}	$9.78x10^{-5\ h}$ (8 day) -	-	$1x10^{-6++}$ (short term) $2x10^{-6++}$ (long term)
Organophosphorus pesticides	Diazinon	*Dm*	$8.6x10^{-4\ a}$	$1.1x10^{-3w}$	$2x10^{-4w}$ (21 day)	-	$1x10^{-5**}$ $1.7x10^{-4+}$
		Dc	$8x10^{-4w}$ (32h)	-	-	-	
		Cd	-	$2.5x10^{-4w}$	-	-	
		Bc	29.22^{b}	-			
			31^{c}	31^{w}	11^{w} (2 day)	-	
	Chlorpyrifos	*Cd*	$1.7x10^{-4w}$	$5x10^{-5\ j}$	-	-	$2x10^{-5*}$ (short term) $2x10^{-6*}$ (long term) $1x10^{-5**}$ $8.3x10^{-5+}$
		Dc	-	$3x10^{-4\ k}$	-	-	
				$5x10^{-4\ l}$	-	-	
		Dg	-	$3x10^{-4\ m}$	-	-	
		Dm	$4.48x10^{-4\ o}$	$7.12x10^{-3\ E}$	-	$5.2\ x10^{-3\ o}$	
		Bc	12^{c}	12^{w}	0.36^{w} (2 day)		
	Methylparathion	*Bc*	29.19^{s}	-	-	-	--
		Bp	10.6^{w}	-	-	-	
		Lq	-	9.49^{r}	6.6^{r} (5 day)	9.4^{r}(45 min)	
		Dm	$3.1x10^{-7\ t}$	$12x10^{-3w}$	-	-	
		Cd		$2.6x10^{-3w}$	-	-	
	Malathion	*Dm*	$4.08x10^{-3\ o}$	$1.6x10^{-3w}$	$3.6x10^{-4w}$ (16 day)	$5.2\ x10^{-3\ o}$	$5x10^{-5**}$
		Dc	-	0.013^{w}	-	-	
		Cd	$3.18x10^{-3w}$	$1.14x10^{-3w}$	-	-	
		Bc	33.72^{u}		-	-	

Dm: Daphnia magna; Dg: Daphnia galeata; Dc: Daphnia carinata; Cd: Ceriodaphnia dubia: Bc: Brachionus calyciflorus; Bp: Brachionus patulus; Lq: Lecane quadridentata

S1: strain 1; S2: strain 2; S3: strain 3

References for toxicological data:[a]: [26], [b]: [52], [c]: [55], [d]: [27], [e]: [28], [f]: [32], [g]: [33], [h]: [34], [i]: [44], [j]: [35], [k]: [36], [l]: [37], [m]: [38], [n]: [41], [o]: [45], [p]: [46], [q]: [43], [r]: [61], [s]: [53], [t]: [62], [u]: [63], [v]: [64], [w]: [65]

References for water quality criteria:*: [66], **: [67], [+] [68], [++][69]

References

1. Obiri-Danso, K., Adonadaga, M. G. and Hogarh, J. N. Effect of agrochemical use on the drinking water quality of Agogo, a tomato growing community in Ashanti Akim, Ghana. Bulletin of Environmental Contamination and Toxicology. 2011. 86:71 – 77. DOI: 10.1007/s00128-010-0183-x
2. Kungolos, A., Emmanouil, C., Tsiridis, V. and Tsiropoulos, N. Evaluation of toxic and interactive toxic effects of three agrochemicals and copper using a battery of microbiotests. Science of the Total Environment. 2009. 407: 4610 – 4615. DOI: 10.1016/j.scitotenv.2009.04.038
3. Kaushik, A., Sharma, H. R., Jain, S., Dawra, J and Kaushik, C. P. Pesticide pollution of River Ghaggar in Haryana, India. Environmental Monitoring and Assessment. 2010. 160: 61 – 69. DOI: 10.1007/s10661-008-0657-z
4. Bhandari, G. An overview of agrochemicals and their effects on environment in Nepal. Applied Ecology and Environmental Sciences. 2014. 2(2): 66 – 73. DOI: 10.12691/aees-2-2-5
5. Dewhurst, I. C. Toxicological assessment of biological pesticides. Toxicology Letters. 2001. 120: 67 – 72. DOI: 10.1016/S0378-4274(01)00308-3
6. Sodhi, G. S. Fundamental Concepts of Environmental Chemistry. 2nd ed. United Kingdom: Alpha Science International Ltd; 2005. 537 p.
7. Konstantinou, I. K., Hela, D. G. and Albanis, T. A. The status of pesticide pollution in surface waters (rivers and lakes) of Greece. Part I. Review on occurrence and levels. Environmental Pollution. 2006. 141: 555 – 570. DOI: 10.1016/j.envpol.2005.07.024
8. Peters, K., Bundschuh, M. and Schäfer, R. B. Review on the effects of toxicants on freshwater ecosystems functions. Environmental Pollution. 2013. 180: 324 – 329. DOI: 10.1016/j.envpol.2013.05.025
9. Bolognesi, C. and Hayashi, M. Review: Micronucleus assay in aquatic animals. Mutagenesis. 2011. 26(1): 205 – 213. DOI: 10.1093/mutage/geq073
10. Cairns, J. Jr. and Pratt, J. R. The scientific basis of bioassays. Hydrobiologia. 1989. 188/189: 5 – 20. DOI: 10.1007/978-94-009-1896-2_2
11. Okamura, H., Omori, M., Luo, R., Aoyama, I. and Liu, D. Application of short-term bioassay guided chemical analysis for water quality of agricultural land run-off. The Science of the Total Environment. 1999. 234: 223 – 231. DOI: 10.1016/S0048-9697(99)00263-6
12. Kokkali, V. and Delft, W. Overview of commercially available bioassays for assessing chemical toxicity in aqueous samples. Trends in Analytical Chemistry. 2014. 61: 133 – 155. DOI: 10.1016/j.trac.2014.08.001

13. Castillo-Morales, G. editor. Ensayos toxicológicos y métodos de evaluación de calidad de aguas. Estandarización, intercalibración, resultados y aplicaciones. 1st ed. IMTA; 2004. 188 p.
14. Kopciuch, R. G., Berecka, B., Bartoszewicz, J. and Buszewski, B. Review: Some considerations about biondicators in environmental monitoring. Polish Journal of Environmental Studies. 2004. 13 (5): 453 – 462
15. Rand, G. M., editor. Fundamentals of Aquatic Toxicology. Effects, Environmental Fate and Risk Assessment. 2nd ed. CRC Press; 1995. 1148 p.
16. Villaroel, M. J., Sancho, E., Ferrando, M. D. and Andreu, E. Acute, chronic and sublethal effects of the herbicide propanil on Daphnia magna. Chemosphere. 2003. 53: 857 – 864. DOI: 10.1016/S0045-6535(03)00546-0
17. Van Leeuwen, C. J. and Vermeire, T. G. Risk Assessment of Chemicals: An Introduction. 2nd ed. Springer; 2007. 686 p. DOI: 10.1007/978/-1-4020-6102-8
18. Farré, M. and Barceló, M. Biosensors for aquatic toxicology evaluation. The Handbook of Environmental Chemistry; 2009. 5J: 115 – 160. DOI: 10.1007/698_2008_13
19. Hoffman, D. J., Barnett, A. R., Burton, G. A. and Cairns, J. Jr., editors. Handbook of Ecotoxicology. 2nd ed. CRC Press; 2003. 1290 p.
20. De Coen, W. M., Janssen, C. R. and Giesy, J. P. Contribution 2. Biomarker applications in ecotoxicology: bridging the gap between toxicology and ecology. New Microbiotests for Routine Toxicity Screening and Biomonitoring. 2000. 13 – 25. DOI: 10.1007/978-1-4615-1289-6_2
21. Connon, R., Dewhurst, R. E., Crane, M. and Callaghan, A. Haem Peroxidase activity in Daphnia magna: A biormarker for sub-lethal toxicity assessment of kerosene-contaminated groundwater. Ecotoxicology. 2003. 12: 387 – 395. DOI: 10.1023/A:1026195621777
22. Pellegri, V., Gorbi, G. and Buschini, A. Comet assay on Daphnia magna in eco-genotoxicity testing. Aquatic Toxicology, 2014. 155: 261 – 268. DOI: 10.1016/j.aquatox.2014.07.002
23. Blaise, C. and Férard, J. F., editors. Overview of contemporary toxicity testing. Small-scale Freshwater Toxicity Investigations. 2005. I: 1 – 68. DOI: 10.1007/1-4020-3120-3_1
24. Santojanni, A., Gorbi, G. and Sartore, F. Prediction of fecundity in chronic toxicity tests son Daphnia magna. Water Research. 1998. 32(10): 3146 – 3156. DOI: 10.1016/S0043-1354(98)00052-9
25. Tatarazako, N. and Oda, S. The water flea Daphnia magna (Crustacea, Cladocera) as a test species for screening and evaluation of chemicals with disrupting effects on crustaceans. Ecotoxicology. 2007. 16: 197 – 203. DOI: 10.1007/s10646-006-0120-2
26. Sánchez, M., Ferrando, M. D., Sancho, E. and Andreu, E. Assessment of the toxicity of a pesticide with a two-generation reproduction test using Daphnia magna.

Comparative Biochemistry and Physiology Part C. 1999. 124: 247 – 252. DOI: 10.1016/S0742-8413(99)00071-7

27. Bettinetti, R., Croce, V., Noe, F., Ponti, B., Quadroni, S. and Galassi, S. Ecotoxicity of pp´DDE to Daphnia magna. Ecotoxicology. 2013. 22: 1255 – 1263. DOI: 10.1007/s10646-013-1113-6
28. Martins, J., Soares, M. L., Saker, M. L., OlivaTeles, L. and Vasconcelos, V. M. Phototactic behaviour in Daphnia magna Straus as an indicator of toxicants in the aquatic environment. Ecotoxicology and Environmental Safety. 2007. 67: 417 – 422. DOI: 10.1016/j.ecoenv.2006.11.003
29. De Meester, L. Short communication. An estimation of the heritability of phototaxis in Daphnia magna Straus. Oecologia. 1989. 78: 142 – 144. DOI: 10.1007/BF00377210
30. Bandala-Sanchez, E., Annesley, S. J. and Fisher, P. R. Short Communication. A phototaxis signaling complex in Dictyostelium discoideum. European Journal of Cell Biology. 2006. 85: 1099 – 1106. DOI: 10.1016/j.ejcb.2006.04.005
31. Yoshizawa, K. and Nogami, S. The first report of phototaxis of fish ectoparasite, Argulus japonicus. Research in Veterinary Science. 2008. 85: 128 – 130. DOI: 10.1016/j.rvsc.2007.09.007
32. Pereira, J. L. and Goncalves, F. Effects of food availability on the acute and chronic toxicity of the insecticide methomyl to Daphnia spp. Science of the Total Environment. 2007. 386: 9 – 20. DOI: 10.1016/j.scitotenv.2007.07.040
33. Toumi, H., Boumaiza, M., Millet, M., Radetski, C. M., Felten, V., Fouque, C. and Férard, J. F. Effects of deltamethrin (pyrethroid insecticide) on growth, reproduction, embryonic development and sex differentiation in two strains of Daphnia magna (Crustacea, Cladocera). Science of the Total Environment. 2013. 458-460: 47 – 53. DOI: 10.1016/j.scitotenv.2013.03.085
34. Shen, M.F., Kumar, A., Ding, S.Y. and Grocke, S. Comparative study on the toxicity of pyrethroids, α-cypermethrin and deltamethrin to Ceriodaphnia dubia. Ecotoxicology and Environmental Safety. 2012. 78: 9 – 13. DOI: 10.1016/j.ecoenv.2011.07.018
35. El-Merhibi, A., Kumar. A. and Smeaton, T. Role of piperonyl butoxide in the toxicity of chlorpyrifos to Ceriodaphnia dubia and Xenopus laevis. Ecotoxicology and Environmental Safety. 2004. 57: 202 – 212. DOI: 10.1016/S0147-6513(03)00082-4
36. Cáceres, T., He, W., Naidu, R. and Megharaj, M. Toxicity of chlorpyrifos and TCP alone and in combination to Daphnia carinata: The influence of microbial degradation. Water Research. 2007. 41: 4497 – 4503. DOI: 10.1016/j.watres.2007.06.025
37. Zalizniak, L. and Nugegoda, D. Effect of sublethal concentrations of chlorpyrifos on three successive generations of Daphnia carinata. Ecotoxicology and Environmental Safety. 2006. 64: 207 – 214. DOI: 10.1016/j.ecoenv.2005.03.015

38. Van Wijngaarden, R., Leeuwangh, P., Lucassen, W. G. H., Romijn, K., Ronday, R., Van der Velde, R. and Willigenburg, W. Acute Toxicity of Chlorpyrifos to Fish, a Newt and Aquatic invertebrates. 1993. 51: 716 – 723. DOI: 10.1007/BF00201650

39. Van den Brink, P. J., Van Donk, E., Gylstra, R., Crum, S. J. H. and Brock, T. C. M. Effects of chronic low concentrations of the pesticides chlorpyrifos and atrazine in indoor freshwater microcosms. Chemosphere. 1995. 31(5): 3181 – 3200. DOI: 10.1016/0045-6535(95)00180-G

40. Guilhermino, L., Lopes, M. C., Carvalho, A. P. and Soares, A. M. V. M. Acetilcholinesterase activity in juveniles of Daphnia magna Straus. Bulletin of Environmental Contamination and Toxicology. 1996. 57: 979 – 985. DOI: 10.1007/s001289900286

41. Liu, H., Yuan, B. and Li, S. Altered quantities and in vivo activities of cholinesterase from Daphnia magna in sub-lethal exposure to organophosphorus insecticides. Ecotoxicology and Environmental Safety. 2012. 80: 118-125. DOI: 10.1016/j.ecoenv.2012.02.014

42. Pereira, J. L., Hill, C. J., Sibly, R. M., Bolshakov, V. N., Goncalves, F., Heckmann, L. H. and Callaghan, A. Gene transcription in Daphnia magna: Effects of acute exposure to a carbam insecticide and an acetanilide herbicide. Aquatic Toxicology. 2010. 97: 268 – 276. DOI: 10.1016/j.aquatox.2009.12.023

43. Jeon, J., Kretschmann, A., Escher, B. I. and Hollender, J. Characterization of acetylcholinesterase inhibition and energy allocation in Daphnia magna exposed to carbaryl. Ecotoxicology and Environmental Safety. 2013. 98: 28 – 35. DOI: 10.1016/j.ecoenv.2013.09.033

44. Toumi, H., Boumaiza, M., Millet, M., Radetski, C. M., Felten, V. and Férard, J. F. Is acetylcholinesterase a biomarker of susceptibility in Daphnia magna (Crustacea, Cladocera) after deltamethrin exposure? Chemosphere. 2015. 120: 351 – 356. DOI: 10.1016/j.chemosphere.2014.07.087

45. Barata, C., Solayan, A. and Porte, C. Role of B-esterases in assessing toxicity of organophoshphorus (Chlorpyrifos, malathion) and carbamate (carbofuran) pesticides to Daphnia magna. Aquatic Toxicology. 2004. 66: 125-139. DOI: 10.1016/j.aquatox.2003.07.004

46. De Coen, W. M., Vangheluwe, M. L. and Janssen, C. R. The use of biomarkers in Daphnia magna toxicity testing. III. Rapid toxicity testing of pure chemicals and sediment pore waters using ingestion and digestive enzyme activity. Chemosphere. 1998. 37(13): 2677 – 2694. DOI: 10.1016/S0045-6535(98)00154-4

47. Belden, J. B. and Lydy, M. J. Impact of atrazine on organophosphate insecticide toxicity. Environmental Toxicology and Chemistry. 2000. 19(9): 2266 – 2274. DOI: 10.1002/etc.5620190917

48. Guilhermino, L., Lopes, M. C., Carvalho, A. P. and Soares, A. M. V. M. Inhibition of acetylcholinesterase activity as effect criterion in acute tests with juvenile Daphnia magna. Chemosphere. 1996. 32 (4): 727 – 738. DOI: 10.1016/0045-6535(95)00360-6

49. Janssen, C. R., Ferrando-Rodrigo, M. D. and Persoone, G. Ecotoxicological studies with the freshwater rotifer Brachionus calyciflorus. I. Conceptual framework and applications. Hydrobiologia. 1993. 255/256: 21 – 32. DOI: 10.1007/978-94-011-1606-0_3

50. Snell, T. W. and Janssen, C. R. Rotifers in ecotoxicology: a review. Hydrobiologia. 1995. 313/314: 231 – 247. DOI: 10.1007/978-94-009-1583-1_32

51. Ma, Q., Xi, Y. L., Zhang, J. Y., Wen, X. L. and Xiang, X. L. Differences in the life table demography among eight geographic populations of Brachionus calyciflorus (Rotifera) from China. Limnologica. 2010. 40: 16 – 22. DOI: 10.1016/j.limno.2009.05.002

52. Fernández-Casalderrey, A., Ferrando, M. D. and Andreu-Moliner, E. Effect of sublethal diazinon concentrations on the demographic parameters of Brachionus calyciflorus Pallas (Rotifera). Bulletin of Environmental Contamination and Toxicology. 1992. 48: 202 – 208. DOI: 10.1007/BF00194372

53. Fernández-Casalderrey, A., Ferrando, M. D. and Andreu-Moliner, E. Chronic toxicity of methylparathion to the rotifer Brachionus calyciflorus fed on Nannochloris oculata and Chlorella pyrenoidosa. Hydrobiologia. 1993. 255/256: 41 – 49. DOI: 10.1007/978-94-011-1606-0_5

54. Ke, L. X., Xi, Y. L., Zha, C. W. and Dong, L. L. Effects of three organophosphorus pesticides on population growth and sexual reproduction of rotifer Brachionus calyciflorus Pallas. Acta Ecologica Sinica. 2009. 29: 182 – 185. DOI: 10.1016/j.chnaes.2009.07.008

55. Juchelka, C. M. and Snell, T. W. Rapid toxicity assessment using rotifer ingestion rate. Archives of Environmental Contamination and Toxicology. 1994. 26: 549 – 554. DOI: 10.1007/BF00214160

56. Rao, T. R. and Sarma, S. S. S. Demographic parameters of Brachionus patulus Muller (Rotifera) exposed to sublethal DDT concentrations at low and high food levels. Hydrobiologia. 1986. 139: 193 – 200. DOI: 10.1007/BF00028292

57. Sarma, S. S. S. and Elías-Guitérrez, M. Rotifers (Rotifera) from four natural water bodies of central Mexico. Limnologia. 1999. 29: 475 – 483. DOI: 10.1016/S0075-9511(99)80054-1

58. Torres-Guzmán, F., Avelar-González, F. J. and Martínez-Rico, R. Implementing Lecane quadridentata acute toxicity tests to assess the toxic effects of selected metals (Al, Fe and Zn). Ecotoxicology and Environmental Safety. 2010. 73: 287-295. DOI: 10.1016/j.ecoenv.2009.10.006

59. Pérez-Legaspi, I. A. and Martínez-Rico, R. Acute toxicity tests on three species of the genus Lecane (Rotifera: Monogononta). Hydrobiologia. 2001. 446/447: 375 – 381. DOI: 10.1023/A:1017531712808

60. Pérez-Legaspi, I. A., Martínez-Rico, R. and Pineda-Rosas, A. Toxicity testing using esterase inhibition as a biomarker in three species of the genus Lecane (Rotifera). Environmental Toxicology and Chemistry. 2002. 21(4): 776-782. DOI: 10.1002/etc.5620210413

61. Pérez-Legaspi, I. A., Quintanar, J. L. and Martínez-Rico, R. Comparing toxicity endpoints on Lecane quadridentata (Rotifera: Monogononta) exposed to two anticholinesterases pesticides. Environmental Toxicology. 2010. 27: 518 – 525. DOI: 10.1002/tox.20668

62. Fernández-Casalderrey, A., Ferrando, M. D and Andreu-Moliner, E. Chronic toxicity of methylparathion to Daphnia magna: effects on survival, reproduction and growth. Bulletin of Environmental Contamination and Toxicology. 1995. 54: 43-49. DOI: 10.1007/BF00196268

63. Fernández-Casalderrey, A., Ferrando, M. D. and Andreu-Moliner, E. Acute toxicity of several pesticides to rotifer (Brachionus calyciflorus). Bulletin of Environmental Contamination and Toxicology. 1992. 48: 14-17. DOI: 10.1007/BF00197477

64. Ferrando, M. D. and Andreu-Moliner. Acute lethal toxicity of some pesticides to Brachionus calyciflorus and Brachionus plicatilis. Bulletin of Environmental Contamination and Toxicology. 1991. 47: 479-484. DOI: 10.1007/BF01702214

65. PAN Pesticides Database - Chemical Toxicity Studies on Aquatic Organisms. Aquatic toxicity. Available from: http://www.pesticideinfo.org/ [Accessed: 2015-01-29]

66. Canadian Council of Ministers of the Environment (CCME). Toxicity database. Available from: http://st-ts.ccme.ca/en/index.html [Accessed: 2015-01-29]

67. National water quality management strategy. Paper No. 4. Australian and New Zealand Guidelines for Fresh and Marine Water Quality. Vol. 2. Aquatic Ecosystems – Rationale and Background Information. Chapter 8. 2000.

68. National Recommended Water Quality Criteria. United States Environmental Protection Agency. Available from: http://water.epa.gov/scitech/swguidance/standards/criteria/current/index.cfm#F [Accessed: 2015-01-29]

69. Fojut, T. L., Mulligan, R. and Tjeerdema, R. S. Water quality criteria report for Cypermethrin. Phase III. Application of the Pesticide Water Quality Criteria Methodology. University of California, Davis. Department of Environmental Toxicology. 2011.

Index

D

E

F

G

Printed at Repro India Ltd.